Name _____ Class _____ Date _____

Skills Worksheet

Directed Reading A

Section: Asking About Life

1. The study of living things is called _____.

IT ALL STARTS WITH A QUESTION

2. The existence of one-celled algae, giant redwood trees, and 40-ton whales illustrates the amazing _____ of life.

3. What is one question you could ask about any living thing?

_____

_____

LIFE SCIENTISTS

4. Who can become a life scientist?

_____

5. Where are some of the places that life scientists work?

_____

_____

6. What determines what a life scientist studies?

_____

WHY ASK QUESTIONS?

7. Life scientists are learning about _____ in order to try to find a cure for this deadly disease.

8. Studying the way humans inherit the code that controls their cells will help scientists learn about diseases such as cystic fibrosis, which may be _____ by children from their parents.

9. What is a major cause of many environmental problems?

_____

10. Why are scientists studying the food and habitat needs of Siberian tigers?

_____

_____

Copyright © by Holt, Rinehart and Winston. All rights reserved.

Holt Science and Technology • The World of Life Science

Name _____ Class _____ Date _____

Skills Worksheet
Directed Reading A

Section: Scientific Methods
WHAT ARE SCIENTIFIC METHODS?

1. The first step is using scientific methods is asking questions. Name two steps that follow.

   _____

   _____

2. Why do scientists vary the order of the steps of scientific methods?

   _____

   _____

ASK A QUESTION

_____ 3. Observations are useful only if they are
   a. important.
   b. accurate.
   c. complicated.
   d. understood.

FORM A HYPOTHESIS

4. A possible explanation or answer to a question is a(n) _____.

5. A hypothesis must be _____ in order to be useful.

6. A statement of cause and effect that can be used to set up a test for a hypothesis is called a(n) _____.

TEST THE HYPOTHESIS

7. What is a controlled experiment?

   _____

   _____

8. What is a variable?

   _____

9. Designing an experiment requires _____.

Name _____ Class _____ Date _____

Directed Reading A *continued*

10. Why do scientists try to test many individuals?

_____

_____

ANALYZE THE RESULTS

11. How might a scientist organize data in order to analyze them?

_____

DRAW CONCLUSIONS

12. Which is more helpful, proving a hypothesis wrong or supporting a hypothesis? Why?

_____

_____

13. Finding an answer to a question often leads to _____.

COMMUNICATE RESULTS

14. What are two reasons that scientists share their results?

_____

_____

Name _____ Class _____ Date _____

Skills Worksheet
Directed Reading A

Section: Scientific Models
TYPES OF SCIENTIFIC MODELS

_____ 1. What is a representation of an object or a system called?
   a. the real thing
   b. a structure
   c. a model
   d. a prediction

_____ 2. Which is an example of a physical model?
   a. an equation
   b. a microscope
   c. a toy rocket
   d. human bones

_____ 3. A Punnett square is an example of
   a. a physical model.
   b. a mathematical model.
   c. a conceptual model.
   d. a representation of an object.

_____ 4. Which of the following is a conceptual model?
   a. the idea that life originated from chemicals
   b. a model human skeleton
   c. $x + 2 = 7$
   d. a plastic human heart

BENEFITS OF MODELS

_____ 5. Models can be used to represent things that are
   a. very small.
   b. very large.
   c. very complicated.
   d. all of the above

BUILDING SCIENTIFIC KNOWLEDGE

6. An explanation that ties together many hypotheses and observations is called a(n) _____.

7. A scientific idea that rarely changes is a scientific _____.

8. What is the difference between scientific theory and scientific law?

   _____

   _____

   _____

9. New scientific ideas take time to develop into _____ or to become accepted as _____ or _____.

Name _____ Class _____ Date _____

Skills Worksheet
Directed Reading A

Section: Tools, Measurement, and Safety

1. What do life scientists use tools for?

   _____

   _____

COMPUTERS AND TECHNOLOGY

2. What is technology?

   _____

   _____

3. When was the first electronic computer built?

   _____

4. How do scientists use computers?

   _____

   _____

TOOLS FOR SEEING

Match the correct description with the correct term. Write the letter in the space provided.

_____ 5. bounces electrons off the surface of a specimen to produce a three-dimensional image

_____ 6. passes electrons through a specimen to produce a flat image

_____ 7. made up of three main parts: a tube with lenses, a stage, and a light

_____ 8. sends electromagnetic waves through the body to create images

a. magnetic resonance imaging
b. scanning electron microscope
c. compound light microscope
d. transmission electron microscope

MEASUREMENT

_____ 9. Many standardized units of measurement were once based on
   a. the weather.
   b. mythology.
   c. parts of the body.
   d. ancient worldwide standards.

Name _____ Class _____ Date _____

Directed Reading A *continued*

10. What are two advantages of using the *International System of Units*?

_____

_____

11. What unit of measurement would a life scientist use to describe the length of an ant? _____

12. A measure of the surface of an object or region is its _____.

13. The units for area are called _____ units.

14. What is the term used to describe the amount of space something takes up or the amount of space it contains?

_____

15. What SI units of measurement are used to determine the volume of liquids and solids?

_____

_____

_____

_____

16. A measure of the amount of matter of an object is its _____.

17. What units of measurement are used to describe the amount of matter, or mass, in an object?

_____

18. The measurement of how hot or cold something is, and an indication of how much energy it contains, is its _____.

19. What are three different units for measuring temperature?

_____

_____

_____

Name _____ Class _____ Date _____

Directed Reading A continued

SAFETY RULES!

20. What are two examples of safety precautions you should follow before beginning an experiment?

_____

_____

Match the symbol with the appropriate label, and write the corresponding letter in the space provided.

_____ 21. hand safety

_____ 22. sharp object

_____ 23. clothing protection

_____ 24. chemical safety

_____ 25. eye protection

_____ 26. electric safety

_____ 27. plant safety

_____ 28. heating safety

_____ 29. animal safety

Name _____ Class _____ Date _____

Skills Worksheet

Directed Reading B

Section: Asking About Life
Circle the letter of the best answer for each question.

1. What is the study of living things called?
   a. technology
   b. life science
   c. investigation
   d. asking questions

IT ALL STARTS WITH A QUESTION

2. What do algae, redwood trees, and whales show?
   a. the diversity of life
   b. life science
   c. lab investigations
   d. asking questions

In Your Own Backyard

3. Which of the following is a life science question you might ask about an organism?
   a. Can that model airplane fly?
   b. What is your dog's name?
   c. How are you feeling today?
   d. Why do leaves change color in the fall?

Name _____ Class _____ Date _____

Directed Reading B *continued*

LIFE SCIENTISTS
Anyone
Circle the letter of the best answer for each question.

4. Who can become a life scientist?

    a. only men

    b. only women

    c. only people who can see

    d. anyone who wants to

Anywhere

5. Life scientists can work

    a. only in a laboratory.

    b. mostly in hospitals.

    c. anywhere they can study living things.

    d. only where there are trees.

Anything

6. What decides what a life scientist studies?

    a. where the person lives

    b. what interests the person

    c. being a man or woman

    d. the time of year

Copyright © by Holt, Rinehart and Winston. All rights reserved.
Holt Science and Technology   The World of Life Science

Name _____ Class _____ Date _____

Directed Reading B *continued*

WHY ASK QUESTIONS?

Read the words in the box. Read the sentences. Fill in each blank with the words or phrase that best completes the sentence.

| diseases | tigers | inherited |

Fighting Diseases

7. Life scientists learn about _____ in order to try to find cures.

Understanding Inherited Diseases

8. Cystic fibrosis is one of many _____ diseases.

Protecting the Environment

9. One life scientist helps the environment by studying Siberian _____.

Name _____ Class _____ Date _____

Skills Worksheet

Directed Reading B

Section: Scientific Methods
WHAT ARE SCIENTIFIC METHODS?

Read the words in the box. Read the sentences. <u>Fill in each blank</u> with the words or phrase that best completes the sentence.

counting	asking questions
accurate	scientific methods

1. The _____ are a series of steps scientists use to solve problems.

2. One step of the scientific methods is _____.

ASK A QUESTION
Make Observations

3. The students made observations by _____ deformed frogs.

Accurate Observations

4. Observations are useful only if they are _____.

Name _____ Class _____ Date _____

Directed Reading B *continued*

FORM A HYPOTHESIS

Read the words in the box. Read the sentences. <u>Fill in each blank</u> with the words or phrase that best completes the sentence.

| hypothesis | if – then |
---

5. A possible answer to a question is

   a(n) _____.

Predictions

6. A prediction is a(n) _____ statement.

TEST THE HYPOTHESIS

| controlled experiment | variable | factor |
---

7. Anything in an experiment that can influence an experiment's outcome is considered a _____.

Under Control

8. An experiment that tests only one factor at a time is a

   _____.

9. The factor that differs between groups in an experiment is the

   _____.

Name _____ Class _____ Date _____

Directed Reading B *continued*

Designing an Experiment
Circle the letter of the best answer for each question.

10. What does designing an experiment require?

  a. planning

  b. factors

  c. many variables

  d. light

Collecting Data

11. Why do scientists try to test many plants or animals?

  a. to be more certain of their data

  b. to get a good hypothesis

  c. to have many variables

  d. to have a big experiment

ANALYZE THE RESULTS

12. What do scientists do before they analyze the results of an experiment?

  a. They organize the data.

  b. They begin a new experiment.

  c. They draw a conclusion.

  d. They write up their results.

DRAW CONCLUSIONS

13. What are scientists deciding when they draw conclusions?

  a. whether to put the data in a graph

  b. which factor is the variable

  c. whether the results support a hypothesis

  d. which group should be the control group

Directed Reading B continued

Is It the Answer?
Circle the letter of the best answer for each question.

14. What is true about finding an answer to a question?

    a. It may begin another investigation.

    b. No more questions can arise.

    c. The question was not good.

    d. The experiment was done wrong.

COMMUNICATE RESULTS

15. What do scientists usually do with their results?

    a. sell them

    b. share them

    c. put them away

    d. destroy them

Name _____ Class _____ Date _____

Skills Worksheet

Directed Reading B

Section: Scientific Models
TYPES OF SCIENTIFIC MODELS

Circle the letter of the best answer for each question.

1. What is a representation of an object or system?
   - **a.** a model
   - **b.** a prediction
   - **c.** an observation
   - **d.** a limitation

2. What is a problem with models?
   - **a.** They are small.
   - **b.** They are not real.
   - **c.** They are on computers.
   - **d.** They may be physical.

3. Which of these is NOT a type of scientific model?
   - **a.** fashion model
   - **b.** conceptual model
   - **c.** mathematical model
   - **d.** physical model

Physical Models

4. Which is a physical model?
   - **a.** an equation
   - **b.** a comparison
   - **c.** a toy rocket
   - **d.** human bones

Name _____ Class _____ Date _____

Directed Reading B *continued*

Mathematical Models
Circle the letter of the best answer for each question.

5. What kind of model is made of numbers and equations?
   a. mathematical model
   b. scientific method
   c. physical model
   d. conceptual model

6. What do scientists use the Punnett square to predict?
   a. how traits are passed to offspring
   b. when traits appear in parents
   c. which mathematical model to use
   d. how complex a model is

Conceptual Models

7. What kind of scientific model is based on systems of ideas?
   a. mathematical model
   b. physical model
   c. computer model
   d. conceptual model

BENEFITS OF MODELS

8. What kinds of things are models used to show?
   a. common things
   b. things that no longer exist
   c. very simple things
   d. things of average size

Directed Reading B *continued*

BUILDING SCIENTIFIC KNOWLEDGE
Scientific Theories
Circle the letter of the best answer for each question.

9. What is an explanation that unites a broad range of facts?
   a. an idea
   b. a theory
   c. an hypothesis
   d. a law

Scientific Laws

10. What kind of scientific idea rarely changes?
    a. an idea
    b. a theory
    c. an hypothesis
    d. a law

Scientific Change

11. Which of the following is true of scientific facts or laws?
    a. Many started as hypotheses.
    b. They never are contradicted.
    c. They are the same thing as theories.
    d. They are conceptual models.

Name _____ Class _____ Date _____

Skills Worksheet

Directed Reading B

Section: Tools, Measurement, and Safety

Read the words in the box. Read the sentences. **Fill in each blank** with the words or phrase that best completes the sentence.

| technology | analyze data |
| tools | computer |

1. Life scientists use various _____ to help them with their work.

COMPUTERS AND TECHNOLOGY

2. The use of machines to meet human needs is called _____.

3. The first _____ was built in 1946.

4. Scientists use computers to _____.

TOOLS FOR SEEING

Read the description. Then <u>draw a line</u> from the dot to the matching word.

5. passes electrons through something to make a 3-D image
6. passes electrons through something to make a flat image
7. is made up of three main parts: a tube with lenses, a stage, and a light
8. sends electromagnetic waves through the body to make images

a. magnetic resonance imaging
b. scanning electron microscope
c. compound light microscope
d. transmission electron microscope

Name _____ Class _____ Date _____

Directed Reading B *continued*

MEASUREMENT

Circle the letter of the best answer for each question.

9. Which is an advantage of the SI system?
   a. is based on grains of wheat
   b. it helps scientists share information
   c. is based on astronomy
   d. it works most of the time

Length

10. Which unit is used for measuring length?
    a. grams (g)
    b. milliliters (mL)
    c. millimeters (mm)
    d. cubic centimeters (cm^3)

Area

11. What is a measure of how much surface an object has?
    a. area
    b. length
    c. micrometers
    d. volume

Volume

12. Which of the following is NOT used to measure volume?
    a. square micrometer
    b. cubic centimeter
    c. milliliter
    d. liter

Name _____ Class _____ Date _____

Directed Reading B continued

Circle the letter of the best answer for each question.

13. What term refers to the amount of space an object takes up?
    a. its length
    b. its area
    c. its volume
    d. its mass

Mass

14. What term refers to the amount of matter in an object?
    a. its length
    b. its area
    c. its volume
    d. its mass

Temperature

15. Which units are part of the International System of Units?
    a. kelvins
    b. ounces
    c. degrees Fahrenheit
    d. degrees

16. What shows how much energy is in matter?
    a. its mass
    b. its temperature
    c. its volume
    d. its length

Name _____ Class _____ Date _____

Directed Reading B *continued*

SAFETY RULES!

Read the description. Then, <u>draw a line</u> from the dot next to each description to the matching picture.

17. hand safety   •   a.
18. eye protection   •   b.
19. plant safety   •   c.
20. electric safety •   d.

Name _____ Class _____ Date _____

Skills Worksheet
Vocabulary and Section Summary

Asking About Life

VOCABULARY

In your own words, write a definition of the following term in the space provided.

1. life science

_____

_____

SECTION SUMMARY

Read the following section summary.

- Science is a process of gathering knowledge about the natural world. Science includes making observations and asking questions about those observations. Life science is the study of living things.
- A variety of people may become life scientists for a variety of reasons.
- Life science can help solve problems such as disease or pollution, and it can be applied to help living things survive.

Name _____ Class _____ Date _____

Skills Worksheet

Vocabulary and Section Summary

Scientific Methods

VOCABULARY

In your own words, write a definition of the following terms in the space provided.

1. scientific methods

   _____

   _____

2. hypothesis

   _____

   _____

3. controlled experiment

   _____

   _____

4. variable

   _____

   _____

SECTION SUMMARY

Read the following section summary.

- Scientific methods are the ways in which scientists follow steps to answer questions and solve problems.
- Any information you gather through your senses is an observation. Observations often lead to the formation of questions and hypotheses.
- A hypothesis is a possible explanation or answer to a question. A well-formed hypothesis is testable by experiment.
- A controlled experiment tests only one factor at a time and consists of a control group and one or more experimental groups.
- After testing a hypothesis, scientists analyze the results and draw conclusions about whether the hypothesis is supported.
- Communicating results allows others to check the results, add to their knowledge, and design new experiments.

Name _____ Class _____ Date _____

Skills Worksheet
Vocabulary and Section Summary

Scientific Models

VOCABULARY

In your own words, write a definition of the following terms in the space provided.

1. model

    _____
    _____

2. theory

    _____
    _____

3. law

    _____
    _____

SECTION SUMMARY

Read the following section summary.

- A model is a representation of an object or system. Models often use familiar things to represent unfamiliar things. Three main types of models are physical, mathematical, and conceptual. Models have limitations but are useful and can be changed based on new evidence.
- Scientific knowledge is built as scientists form and revise scientific hypotheses, models, theories, and laws.

Name _____ Class _____ Date _____

Skills Worksheet
Vocabulary and Section Summary

Tools, Measurement, and Safety
VOCABULARY
In your own words, write a definition of the following terms in the space provided.

1. technology
   _____
   _____

2. compound light microscope
   _____
   _____

3. electron microscope
   _____
   _____

4. area
   _____
   _____

5. volume
   _____
   _____

6. mass
   _____
   _____

7. temperature
   _____
   _____

Vocabulary and Section Summary continued

SECTION SUMMARY
Read the following section summary.

- Life scientists use computers to collect, store, organize, analyze, and share data.
- Life scientists commonly use light microscopes and electron microscopes to make observations of things that are too small to be seen without help. Electromagnetic waves are also used in other ways to create images.
- The International System of Units (SI) is a simple and reliable system of measurement that is used by most scientists.

Name _____ Class _____ Date _____

Skills Worksheet
Section Review

Asking About Life
USING KEY TERMS

1. In your own words, write a definition for the term *life science*.

   _____

UNDERSTANDING KEY IDEAS

_____ 2. Life scientists may study any of the following EXCEPT
   a. things that were once living.
   b. environmental problems.
   c. stars in outer space.
   d. diseases that are not inherited by humans.

3. What is the importance of asking questions in life science?

   _____

   _____

   _____

4. Where do life scientists work? What do life scientists study?

   _____

   _____

   _____

MATH SKILLS

5. Students in a science class collected 50 frogs from a pond and found that 15 of these frogs had deformities. What percentage of the frogs had deformities? Show your work below.

Section Review continued

CRITICAL THINKING

6. Identifying Relationships Make a list of five things you do or deal with daily. Give an example of how life science might relate to each of these things.

_____

7. Applying Concepts Look at Figure 5. Propose five questions about what you see. Share one of your questions with your classmates.

_____

Name _____ Class _____ Date _____

Skills Worksheet

Section Review

Scientific Methods

USING KEY TERMS

1. Use the following terms in the same sentence: *hypothesis*, *controlled experiment*, and *variable*.

   _____

   _____

   _____

UNDERSTANDING KEY IDEAS

_____ 2. The steps of scientific methods
   a. are exactly the same in every investigation.
   b. must always be used in the same order.
   c. are not always used in the same order.
   d. always end with a conclusion.

3. What are the essential parts of a controlled experiment?

   _____

   _____

4. What causes scientific knowledge to change?

   _____

   _____

MATH SKILLS

5. Calculate the average of the following values: 4, 5, 6, 6, 9. Show your work below.

Name _____ Class _____ Date _____

Section Review *continued*

CRITICAL THINKING

6. Analyzing Methods Why was UV light chosen to be the variable in the frog experiment?

_____
_____
_____

7. Analyzing Processes Why are there many ways to follow the steps of scientific methods?

_____
_____
_____

8. Making Inferences Why might two scientists working on the same problem draw different conclusions?

_____
_____
_____

9. Identifying Bias Investigations often begin with observation. How does observation limit what scientists can study?

_____
_____
_____
_____

Section Review continued

INTERPRETING GRAPHICS

10. The table below shows how long it takes for one bacterium to divide and become two bacteria. Plot this information on a graph, with temperature on the x-axis and the time to double on the y-axis. Do not graph values for which there is no growth. What temperature allows the bacteria to multiply most quickly?

Temperature (°C)	Time to double (minutes)
10	130
20	60
25	40
30	29
37	17
40	19
45	32
50	no growth

Name _____ Class _____ Date _____

Skills Worksheet
Section Review

Scientific Models
USING KEY TERMS

In each of the following sentences, replace the incorrect term with the correct term from the word bank.

    theory   law

1. A law is an explanation that matches many hypotheses but may still change.

_____

2. A model tells you exactly what to expect in certain situations.

_____

UNDERSTANDING KEY IDEAS

_____ 3. A limitation of models is that
    a. they are large enough to see.
    b. they do not act exactly like the things that they model.
    c. they are smaller than the things that they model.
    d. they model unfamiliar things.

4. What are three types of models? Give an example of each type.

_____

_____

_____

5. Compare how scientists use theories with how they use laws.

_____

_____

_____

_____

_____

_____

Section Review *continued*

MATH SKILLS

6. If Jerry is 2.1 m tall, how tall is a scale model of Jerry that is 10% of his size? Show your work below.

CRITICAL THINKING

7. **Applying Concepts** You and a friend are making a three-dimensional model of an extinct plant. Describe some of the potential uses for your model. What are some limitations of your model?

_____

_____

_____

_____

_____

_____

_____

_____

_____

Name _____ Class _____ Date _____

Skills Worksheet

Section Review

Tools, Measurement, and Safety

USING KEY TERMS

Complete each of the following sentences by choosing the correct term from the word bank.

    mass   area
    volume   temperature

1. The measure of the surface of an object is called _____.

2. Life scientists use kilograms when measuring an object's _____.

3. The _____ of a liquid is usually described in liters.

UNDERSTANDING KEY IDEAS

_____ 4. SI units are
    a. always based on standardized measurements of body parts.
    b. almost always based on the number 10.
    c. used only to measure length.
    d. used only in France.

5. How is temperature related to energy?

_____

_____

6. If you were going to measure the mass of a fly, which SI unit would be most appropriate?

_____

_____

Name _____ Class _____ Date _____

Section Review *continued*

MATH SKILLS

7. Convert 3.0 L into cubic centimeters. Show your work below.

8. Calculate the volume of a textbook that is 28.5 cm long, 22 cm wide, and 3.5 cm thick. Show your work below.

CRITICAL THINKING

9. Making Inferences The mite shown in your textbook is about 500 µm long in real life. What tool was probably used to produce this?

_____

_____

_____

_____

10. Applying Concepts Give an example of what could happen if you do not follow safety rules.

_____

_____

_____

_____

_____

Name _____ Class _____ Date _____

Skills Worksheet

Chapter Review

USING KEY TERMS

1. Use the following terms in the same sentence: *life science* and *scientific methods*.

   _____
   _____

2. Use the following terms in the same sentence: *controlled experiment* and *variable*.

   _____
   _____

For each pair of terms, explain how the meanings of the terms differ.

3. *theory* and *hypothesis*

   _____
   _____
   _____

4. *compound light microscope* and *electron microscope*

   _____
   _____
   _____

5. *area* and *volume*

   _____
   _____
   _____

UNDERSTANDING KEY IDEAS

Multiple Choice

_____ 6. The steps of scientific methods
   a. must all be used in every scientific investigation.
   b. must always be used in the same order.
   c. often start with a question.
   d. always result in the development of a theory.

Chapter Review continued

_____ **7.** In a controlled experiment,
   a. a control group is compared with one or more experimental groups.
   b. there are at least two variables.
   c. all factors should be different.
   d. a variable is not needed.

_____ **8.** Which of the following tools is best for measuring 100 mL of water?
   a. 10 mL graduated cylinder
   b. 150 mL graduated cylinder
   c. 250 mL beaker
   d. 500 mL beaker

_____ **9.** Which of the following is NOT an SI unit?
   a. meter   **c.** liter
   b. foot   **d.** kilogram

_____ **10.** A pencil is 14 cm long. How many millimeters long is it?
   a. 1.4 mm   **c.** 1,400 mm
   b. 140 mm   **d.** 1,400,000 mm

_____ **11.** The directions for a lab include the safety icons shown below. These icons mean that

   a. you should be careful.
   b. you are going into the laboratory.
   c. you should wash your hands first.
   d. you should wear safety goggles, a lab apron, and gloves during the lab.

Short Answer

12. List three ways that science is beneficial to living things.

_____

_____

_____

13. Why do hypotheses need to be testable?

_____

_____

Chapter Review *continued*

14. Give an example of how a life scientist might use computers and technology.

15. List three types of models, and give an example of each.

16. What are some advantages and limitations of models?

17. Which SI units can be used to describe the volume of an object? Which SI units can be used to describe the mass of an object?

18. In a controlled experiment, why should there be several individuals in the control group and in each of the experimental groups?

Name _____ Class _____ Date _____

Chapter Review *continued*

CRITICAL THINKING

19. Concept Mapping Use the following terms to create a concept map: *observations, predictions, questions, controlled experiments, variable,* and *hypothesis.*

20. Making Inferences Investigations often begin with observation. What limits are there to the observations that scientists can make?

21. Forming Hypotheses A scientist who studies mice observes that on the day the mice are fed vitamins with their meals, they perform better in mazes. What hypothesis would you form to explain this phenomenon? Write a testable prediction based on your hypothesis.

Name _____ Class _____ Date _____

Chapter Review continued

INTERPRETING GRAPHICS

The pictures below show how an egg can be measured by using a beaker and water. Use the pictures to answer the questions that follow.

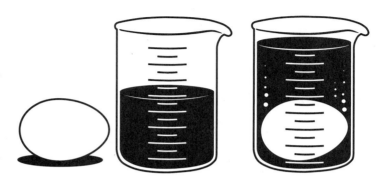

Before: 125 mL   After: 200 mL

_____ 22. What kind of measurement is being taken?
  a. area
  b. length
  c. mass
  d. volume

_____ 23. Which of the following is an accurate measurement of the egg in the picture?
  a. 75 cm³
  b. 125 cm³
  c. 125 mL
  d. 200 mL

24. Make a double line graph from the data in the following table.

Number of Frogs		
Date	Normal	Deformed
1995	25	0
1996	21	0
1997	19	1
1998	20	2
1999	17	3
2000	20	5

Name _____ Class _____ Date _____

Skills Worksheet

Reinforcement

The Mystery of the Bubbling Top
Complete this worksheet after you finish reading the section "Scientific Methods."

Use the materials below to conduct the activity. Then answer the questions that follow.

MATERIALS
- small, empty, plastic soda bottle
- cold water
- plastic or plastic-foam disposable plate
- scissors
- hot water
- beaker or other container large enough to hold the soda bottle

PROCEDURE
1. Fill the empty bottle halfway with cold water.
2. Cut a quarter-sized disk from the plastic plate.
3. Moisten the plastic disk, and place it on top of the bottle's neck.
4. Pour hot water into the beaker until it is about one-quarter full.
5. Carefully place the bottle inside the beaker.
6. What happened to the plastic disk?

_____

_____

_____

_____

YOU JUST MADE OBSERVATIONS.
7. Why do you think the plastic disk did that? Brainstorm for as many answers as possible. Then put a star next to the explanation you consider most reasonable.

_____

_____

_____

_____

Copyright © by Holt, Rinehart and Winston. All rights reserved.

Holt Science and Technology   The World of Life Science

Name _____ Class _____ Date _____

Reinforcement continued

YOU JUST FORMED A HYPOTHESIS.

8. How could you test your hypothesis? Outline an experiment you could conduct.

_____
_____
_____
_____

9. Conduct your experiment. What happened?

_____
_____
_____
_____

YOU JUST TESTED YOUR HYPOTHESIS.

10. How do you explain the results of your experiment?

_____
_____
_____
_____

YOU JUST ANALYZED THE RESULTS OF YOUR EXPERIMENT.

11. Do the results of your experiment match your hypothesis? Explain.

_____
_____
_____

12. Do you need to conduct more experiments to find out if your hypothesis is correct? Why or why not?

_____
_____
_____

You just drew conclusions. Congratulations! You have just finished several steps of the scientific methods! Share your results with your classmates.

Name _____ Class _____ Date _____

Skills Worksheet
Critical Thinking

The Case of the Bulge

Dear J. Q. Public,

We are writing to inform you of a recent finding by our food safety inspector.

At about the same time every year for the last few years, cans of certain foods have bulged in local stores and homes. No one has been able to determine exactly why this happens. We have found that the cans bulge only during summer months and that only locally canned food is affected. Over the last few years, some local canneries have been using a different canning process, which requires less time and lower temperatures. We are uncertain whether there is any connection between the new canning process and the bulging cans.

We are concerned for the safety of the citizens. Bulging cans may contain bacteria called *Clostridium botulinum*, which can cause a type of food poisoning called botulism. *Clostridium botulinum* thrives in warm environments and causes cans to bulge as the bacteria grow. *Clostridium botulinum* may be found in normal or bulging cans, depending on the growth stage.

We would like to solve this case as quickly and efficiently as possible. If you have any clues that will help us, please let us know. Thank you for your time.

City Health Department

Use the scientific methods to investigate the case of the bulge.

STATE THE PROBLEM

1. In your own words, state the central problem that the City Health Department faces.

_____

_____

_____

COMPREHENDING IDEAS

2. What additional information could the City Health Department provide that would help you investigate the problem?

_____

_____

_____

Copyright © by Holt, Rinehart and Winston. All rights reserved.

Holt Science and Technology   The World of Life Science

Name _____ Class _____ Date _____

Critical Thinking continued

FORM A HYPOTHESIS

3. Form a hypothesis that explains the cause of the bulging cans.

_____

_____

_____

TEST THE HYPOTHESIS

4. Design and outline an experiment that the City Health Department could perform to test your hypothesis. Be sure to use a control group and an experimental group in your experiment plan.

_____

_____

_____

_____

_____

_____

_____

COMMUNICATE RESULTS

5. Write a reply letter to the City Health Department. Share your hypothesis and the steps you would take to solve this problem. Include the results that would prove that your hypothesis is correct.

_____

_____

_____

_____

_____

_____

_____

Name _____ Class _____ Date _____

Assessment

Section Quiz

Section: Asking About Life
Write the letter of the correct answer in the space provided.

_____ 1. If you make an observation of a living thing and then ask a question about what you observed, you are
   a. noticing the diversity of life.
   b. behaving like a life scientist.
   c. solving a problem.
   d. learning how to protect the environment.

_____ 2. For every organism that has ever lived,
   a. there is only one question to ask.
   b. many questions could be asked.
   c. every question has already been asked.
   d. every question has already been answered.

_____ 3. Which of the following people is LEAST suited to being a life scientist?
   a. someone who likes sports
   b. someone who goes to school
   c. someone who is very curious
   d. someone who has no interest in organisms

_____ 4. A life scientist is LEAST likely to be found working
   a. in a laboratory.
   b. in a hospital.
   c. in an art museum.
   d. at the bottom of the ocean.

_____ 5. Which of the following is NOT one of the ways in which the work of a life scientist is beneficial?
   a. helping to fight diseases
   b. finding out about weather patterns
   c. studying environmental problems on Earth
   d. studying how humans inherit the code that controls their cells

_____ 6. Questions that life scientists today are trying to answer include all of the following EXCEPT
   a. the part of a person's inherited information responsible for certain inherited diseases.
   b. how the human body responds to space travel.
   c. how shells have changed over time.
   d. the age of the oldest star

Name _____ Class _____ Date _____

Assessment

Section Quiz

Section: Scientific Methods

Use the information in the graph to answer questions 1-3. Write the letter of the correct answer in the space provided.

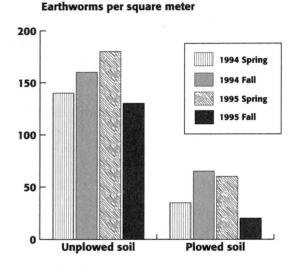

_____ 1. What question did the scientists who collected this data want to answer?
   a. Are there more earthworms in the soil in the spring or in the fall?
   b. What is the effect of plowing soil on the number of earthworms?
   c. How is the size of earthworms affected by the seasons?
   d. Does plowing soil affect how fast earthworms grow?

_____ 2. Where and when were the most earthworms found?
   a. unplowed soil, spring 1995
   b. unplowed soil, fall 1994
   c. unplowed soil, fall 1995
   d. plowed soil, spring 1994

_____ 3. What do the data in this graph show?
   a. Unplowed soil has more earthworms than plowed soil.
   b. Plowed soil has more earthworms than unplowed soil.
   c. Plowing of soil has no effect on the number of earthworms.
   d. The number of earthworms cannot be predicted.

Write the letter of the correct answer in the space provided.

_____ 4. A hypothesis is
   a. a fact.
   b. a type of question.
   c. a possible answer to a question.
   d. an experiment.

_____ 5. A controlled experiment
   a. tests several different factors at one time.
   b. has several control groups.
   c. has more control groups than experimental groups.
   d. has only one variable.

Name _____ Class _____ Date _____

Assessment

Section Quiz

Section: Scientific Models

Match the correct description with the correct term. Write the letter in the space provided. A term may be used more than once.

_____ 1. a Punnet square

_____ 2. a toy train

_____ 3. the idea that life originated from chemicals

_____ 4. a method of classifying the behavior of animals

_____ 5. a dinosaur sculpture in a museum

a. physical model
b. mathematical model
c. conceptual model

Write the letter of the correct answer in the space provided.

_____ 6. Which of the following is a limitation of models?
   a. They help explain how something works.
   b. They help describe how something is structured.
   c. They are different from the real things they are modeling.
   d. They are used to make predictions.

_____ 7. What is the advantage of creating a model of a dinosaur?
   a. Dinosaurs are too large to easily study.
   b. Real dinosaurs cannot be studied because they died out long ago.
   c. Dinosaurs are too complicated to study without a model.
   d. The model is safer to work with.

_____ 8. A unifying explanation for a broad range of observations, facts, and tested hypotheses is called a
   a. theory.
   b. law.
   c. hypothesis.
   d. conclusion.

_____ 9. Life science has few scientific laws because
   a. people don't want them.
   b. life scientists have not done enough experiments.
   c. living organisms are so complex.
   d. scientists need to build more models.

_____ 10. Many scientific laws
   a. have always been known to be true.
   b. are easily contradicted by new experiments.
   c. often get broken.
   d. may have started off as hypotheses or theories.

Copyright © by Holt, Rinehart and Winston. All rights reserved.
Holt Science and Technology   The World of Life Science

Name _____ Class _____ Date _____

Assessment
Section Quiz

Section: Tools, Measurement, and Safety

Match the correct description with the correct term. Write the letter in the space provided. A term may be used more than once.

_____ 1. may be measured in nanometers

_____ 2. the amount of space something takes up

_____ 3. a measure of the size of a surface or region

_____ 4. tells how much energy is in matter

_____ 5. may be described in grams or kilograms

_____ 6. SI units are kelvins

_____ 7. units for solids are cubic

a. area
b. volume
c. mass
d. temperature
e. length

Write the letter of the correct answer in the space provided.

_____ 8. Which technology would be used to view the surface of a tiny living organism?
  a. transmission electron microscope
  b. scanning electron microscope
  c. compound light microscope
  d. computerized tomography scan

_____ 9. Which of the following tools might a scientist use to solve a complex mathematical problem?
  a. compound light microscope
  b. computer
  c. magnetic resonance imaging
  d. hypothesis

_____ 10. Which technology would be used to view a person's internal organs?
  a. magnetic resonance imagery
  b. scanning electron microscope
  c. electronic computer
  d. compound light microscope

Name _____ Class _____ Date _____

Assessment
Chapter Test A

The World of Life Science
MATCHING
Match the correct definition with the correct term. Write the letter in the space provided.

_____ 1. a possible explanation for observations

_____ 2. the amount of space something occupies

_____ 3. used to magnify a living specimen

_____ 4. set of related hypotheses supported by evidence

_____ 5. the use of knowledge, tools, and materials to solve problems and accomplish tasks

_____ 6. the study of living things

_____ 7. a series of steps followed by scientists to solve problems

_____ 8. used to produce clear and detailed images of nonliving specimens

_____ 9. the amount of matter in an object

_____ 10. a summary of many experimental results and observations

a. technology
b. volume
c. electron microscope
d. scientific methods
e. life science
f. mass
g. theory
h. law
i. hypothesis
j. compound light microscope

MULTIPLE CHOICE
Write the letter of the correct answer in the space provided.

_____ 11. In which of the following areas of study might you find a life scientist at work?
   a. discovering ways to improve computer-operated robots
   b. studying how wasps could be used to control fire ant populations
   c. studying the relationship between El Niño and increased flooding
   d. researching the paths of comets and meteors in space

Name _____ Class _____ Date _____

Chapter Test A *continued*

_____ 12. A scientist who wants to study the possible side effects of a new medicine would probably
   a. give each experimental group the same dose of medicine.
   b. simply ask the subjects about the medicine's effects.
   c. include a control group that received no medicine.
   d. use different numbers of subjects in each treatment group.

_____ 13. Which tool would a life scientist use to obtain a detailed image of the nerves that branch off a person's spinal cord?
   a. a scanning electron microscope
   b. a compound light microscope
   c. a transmission electron microscope
   d. an MRI

_____ 14. An advantage of using the International System of Units is that it
   a. is based on measurements of common body parts.
   b. includes easily understood units of measure, such as inches.
   c. gives scientists a common way to share their results.
   d. uses different units to measure mass and weight.

_____ 15. In a scientific experiment, a hypothesis that cannot be tested is always considered to be
   a. incorrect.
   b. illogical.
   c. not useful.
   d. a theory.

_____ 16. Grams are the most useful unit of measurement in determining the mass of a
   a. truck.
   b. medium-sized apple.
   c. grain of rice.
   d. hippopotamus.

Chapter Test A continued

MATCHING

Match the correct definition with the correct term. Write the letter in the space provided.

_____ 17. Write an article describing what you learned about the ant population on school grounds.

_____ 18. Last week there were no ants near the front door of our school. Now there is a large colony. Where did the colony come from?

_____ 19. I think someone released ants from their ant farm near the front door of our school.

_____ 20. There are three ant colonies on the school grounds. Four of the 10 residents who live near the school also have ant colonies in their yards. None of the residents has ever owned an ant farm. None of the students surveyed had information about where the ants came from.

_____ 21. Evidence seems to indicate that our rivals, the Hornets, placed the ant colony on our school grounds.

_____ 22. I am examining the school grounds and surveying students and nearby residents for information about where the ants came from.

a. Ask a question.
b. Form a hypothesis.
c. Test the hypothesis.
d. Analyze the results.
e. Draw conclusions.
f. Communicate results

Chapter Test A *continued*

MULTIPLE CHOICE

Write the letter of the correct answer in the space provided.

_____ 23. A scientific model
   a. can be a kind of hypothesis.
   b. is usually used to represent something simple.
   c. is usually concrete.
   d. is used to explain observations.

Use the graph below to answer question 24. Write the letter or the correct answer in the space provided.

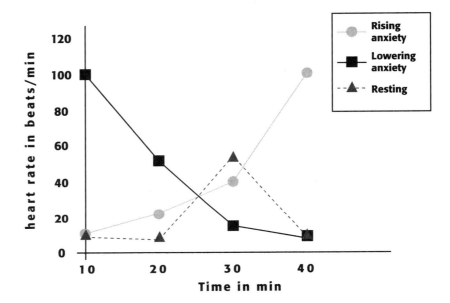

_____ 24. The most likely conclusion to be drawn from the experimental results shown in the graph is that
   a. the heart rate is never stable.
   b. only emotions affect the heart rate.
   c. the heart rate changes only when emotions change.
   d. the heart rate increases with rising anxiety.

Name _____ Class _____ Date _____

Assessment

Chapter Test B

The World of Life Science

USING KEY TERMS

Use the terms from the following list to complete the sentences below. Each term may be used only once. Some terms may not be used.

mass   electron microscope   scientific methods
theory   volume   technology
hypothesis   compound light microscope

1. Scientists often use experiments to test a _____, which is a possible explanation for observations.

2. The _____ of something is defined as the amount of space it occupies.

3. A life scientist would use a(n) _____ to magnify a living specimen.

4. A set of related hypotheses that are supported by evidence may become accepted as a _____.

5. The use of knowledge, tools, and materials to solve problems and accomplish tasks is known as _____.

6. Using _____, scientists follow steps to answer questions and solve problems.

UNDERSTANDING KEY IDEAS

Write the letter of the correct answer in the space provided.

_____ 7. In which of the following areas of study might you find a life scientist at work?
   a. discovering ways to improve computer programs
   b. studying the impact of non-native plants on marshes
   c. trying to develop a warning system for tornadoes
   d. researching the composition of asteroids in space

_____ 8. A scientist who wants to study the effects of a new fertilizer on plants would probably
   a. give each experimental group the same amount of the fertilizer.
   b. not worry about measuring the amount of fertilizer used.
   c. include a control group that received no fertilizer.
   d. use different numbers of plants in each treatment group.

Name _____ Class _____ Date _____

Chapter Test B *continued*

_____ **9.** Which tool would a life scientist use to obtain a detailed image of the blood vessels in a person's leg?
   a. a scanning electron microscope
   b. a compound light microscope
   c. a transmission electron microscope
   d. an MRI

_____ **10.** What is one advantage of using the International System of Units?
   a. It is based on measurements of common body parts.
   b. It includes easily understood units of measure, such as pounds.
   c. Almost all units are based on the number 10.
   d. It was developed in France.

_____ **11.** A 100 kg object contains the same amount of a matter as a
   a. 1000 g object.
   b. 100,000 g object.
   c. 10,000 g object.
   d. 1,000,000 g object.

12. Why is the rejection of a hypothesis helpful to scientists?

_____
_____
_____
_____

13. What is the purpose of the control group in a controlled experiment?

_____
_____
_____
_____

14. What are the advantages and disadvantages of using a model?

_____
_____
_____
_____

Chapter Test B continued

CRITICAL THINKING

15. A scientist forms the following hypothesis: Classical music improves a person's ability to study. What would you expect to find if the scientist's hypothesis were true?

_____

_____

_____

16. How could you demonstrate that 1 mL and 1 cm^3 represent equal quantities?

_____

_____

_____

17. A scientist collected the following data from an experiment on flowering plants. Analyze the scientist's results and determine what conclusions could be drawn.

Flowering of Two Plants

	Plant Group	Number of hours of light		
		10	12	16
Average number of flowers per plant	#1 Plant A	0	2	15
	#2 Plant A	0	1	18
	#3 Plant B	10	4	0
	#4 Plant B	14	5	0

_____

_____

_____

_____

_____

CONCEPT MAPPING

18. Use the following terms to complete the concept map below:

controlled experiments observations new questions
technology hypothesis

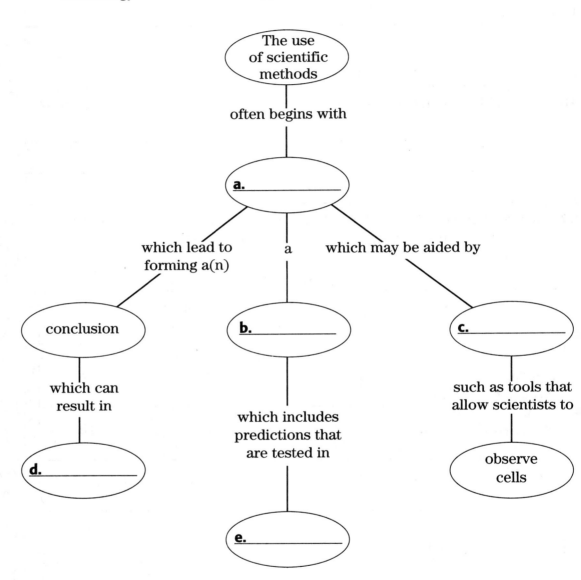

Name _____ Class _____ Date _____

Assessment
Chapter Test C

The World of Life Science
MULTIPLE CHOICE
Circle the letter of the best answer for each question.

1. Which work might a life scientist do?

    a. Build robots.

    b. Study Siberian tigers.

    c. Study what affects flooding.

    d. Research comets in space.

2. Which tool has a tube with lenses, a stage, and a light?

    a. a scanning electron microscope

    b. a compound light microscope

    c. a transmission electron microscope

    d. an MRI

3. Which units are part of the International System of Units?

    a. inches

    b. milliliter

    c. pounds

    d. degrees Fahrenheit

4. Which of these is NOT a type of scientific model?

    a. mathematical model

    b. physical model

    c. fashion model

    d. conceptual model

Chapter Test C continued

MULTIPLE CHOICE
Circle the letter of the best answer for each question.

5. Which is a step of the scientific methods?
   a. asking questions
   b. stating a theory
   c. using technology
   d. building a microscope

6. Which of the following differs between groups in a controlled experiment?
   a. a test
   b. a prediction
   c. a variable
   d. a hypothesis

7. Which is a physical model?
   a. a comparison
   b. an equation
   c. a toy airplane
   d. a graph

8. What term refers to the amount of space an object takes up?
   a. volume
   b. mass
   c. area
   d. length

9. Which describes a compound light microscope?
   a. It passes electrons through something to make a 3-D image.
   b. It passes electrons through something to make a flat image.
   c. It sends electromagnetic waves through the body to make images.
   d. It is made up of three main parts: a tube with lenses, a stage, and a light.

Name _____ Class _____ Date _____

Chapter Test C continued

MATCHING

Read the description. Then, <u>draw a line</u> from the dot next to each description to the matching word.

10. a possible explanation for observations •

11. the amount of space something occupies •

    a. volume

    b. hypothesis

12. the amount of matter in an object •

    c. mass

    d. theory

13. an explanation that unites a broad range of facts •

---

14. the use of machines to meet human needs •

    a. life science

15. the study of living things •

    b. technology

16. a series of steps followed by scientists to solve problems •

    c. scientific methods

17. a kind of scientific idea that rarely changes •

    d. law

Chapter Test C continued

FILL-IN-THE-BLANK

Read the words in the box. Read the sentences. <u>Fill in each blank</u> with the word or phrase that best completes the sentence.

model	scientific methods
life science	controlled experiment

18. A _____ is a representation of an object or a system.

19. The steps scientists use to answer questions are called _____.

20. A _____ tests only one factor at a time.

21. The study of living things is _____.

Name _____ Class _____ Date _____

Assessment

SKILLS PRACTICE

Performance-Based Assessment

OBJECTIVE

Using the scientific method, you will find out why pennies minted in different years have different masses.

KNOW THE SCORE!

As you work through the activity, keep in mind that you will be earning a grade for the following:

- how well you work with materials and equipment (20%)
- how well you formulate hypotheses (20%)
- how well you use the scientific method to prove or disprove your hypotheses (40%)
- how well you record the results of your experiments (20%)

Using Scientific Methods

ASK A QUESTION

Why are pennies from some years heavier than pennies from others?

MATERIALS AND EQUIPMENT

- 12 pennies: 4 from the 1970s, 4 from the 1980s, and 4 from the 1990s
- metric balance
- 50 mL graduated cylinder
- water

SAFETY INFORMATION

- Be careful with glass objects.
- Wipe up all spills immediately to avoid slipping.

FORM A HYPOTHESIS

1. Form a hypothesis to answer the question.

_____

_____

Name _____ Class _____ Date _____

Performance-Based Assessment continued

TEST THE HYPOTHESIS

2. Arrange the pennies in chronological order, from oldest to newest. Starting with the oldest penny, find the mass of each penny and record each mass next to the year in the data chart provided.

3. Fill the graduated cylinder about half full with water. Record the exact volume in the first row of the data chart.

4. Place the first penny in the cylinder and record the new water level in the data chart.

5. Subtract the water level before the penny was added from the water level after the penny was added. This gives you the volume of the penny. Then calculate the density of the penny by dividing the mass by the volume. Record this value in the chart.

6. Repeat this procedure for each of the remaining pennies, using the water level from the previous step as the new starting level. Do not remove any pennies from the cylinder.

Data Collection Table

Year	Mass (g)	Water level before penny is added (mL)	Water level after penny (mL)	Difference in water levels	Density (g/mL)

ANALYZE THE RESULTS

7. What does this experiment tell you about the volume of pennies?

_____

_____

Name _____ Class _____ Date _____

Performance-Based Assessment continued

8. How has the density of pennies changed over the past 30 years?

_____
_____
_____
_____

9. Why might the density have changed?

_____
_____
_____
_____

Name _____ Class _____ Date _____

Assessment
Standardized Test Preparation

READING

Read each of the following passages below. Then, answer the questions that follow the passage.

Passage 1 Zoology is the study of animals. Zoology dates back more than 2,300 years, to ancient Greece. There, the philosopher Aristotle observed and theorized about animal behavior. About 200 years later, Galen, a Greek physician, began dissecting and experimenting with animals. However, there were few advances in zoology until the 1700s and 1800s. During this period, the Swedish naturalist Carolus Linnaeus developed a classification system for plants and animals, and British naturalist Charles Darwin published his theory of evolution by natural selection.

_____ 1. According to the passage, when did major advances in Zoology begin?
 A About 2,300 years ago
 B About 2,100 years ago
 C During the 1700s and 1800s
 D Only during recent history

_____ 2. Which of the following is a possible meaning of the word *naturalist*, as used in the passage?
 F a scientist who studies plants and animals
 G a scientist who studies animals
 H a scientist who studies theory
 I a scientist who studies animal behavior

_____ 3. Which of the following is the **best** title for this passage?
 A Greek Zoology
 B Modern Zoology
 C The Origins of Zoology
 D Zoology in the 1700s and 1800s

Copyright © by Holt, Rinehart and Winston. All rights reserved.

Holt Science and Technology   The World of Life Science

Standardized Test Preparation continued

Passage 2 When looking for answers to a problem, scientists build on existing knowledge. For example, scientists have wondered if there is some relationship between Earth's core and Earth's magnetic field. To form a hypothesis, scientists started with what they knew: Earth has a dense, solid inner core and a molten outer core. Scientists then created a computer model to simulate how Earth's magnetic field might be generated.

They tried different things with their model until the model produced a magnetic field that matched that of the real Earth. The model predicted that Earth's inner core spins in the same direction as the rest of the Earth, but the inner core spins slightly faster than Earth's surface. If the hypothesis is correct, it might explain how Earth's magnetic field is produced. Although scientists cannot reach the Earth's core to examine it directly, they can test whether other observations match what is predicted by their hypothesis.

_____ 1. What does the word *model* refer to in this passage?
   A a giant plastic globe
   B a representation of the Earth created on a computer
   C a computer terminal
   D a technology used to drill into the Earth's core

_____ 2. Which of the following is the **best** summary of the passage?
   F Scientists can use models to help them answer difficult and complex questions.
   G Scientists have discovered the source of Earth's magnetic field.
   H The spinning of Earth's molten inner core causes Earth's magnetic field.
   I Scientists make a model of a problem and then ask questions about the problem.

Name _____ Class _____ Date _____

Standardized Test Preparation continued

INTERPRETING GRAPHICS

The table below shows the plans for an experiment in which bees will be observed visiting flowers. Use the table to answer the questions that follow.

Bee Experiment				
Group	Type of bee	Time of day	Type of plant	Flower color
#1	Honeybee	9:00 A.M.–10:00 A.M.	Portland rose	red
#2	Honeybee	9:00 A.M.–10:00 A.M.	Portland rose	yellow
#3	Honeybee	9:00 A.M.–10:00 A.M.	Portland rose	white
#4	Honeybee	9:00 A.M.–10:00 A.M.	Portland rose	pink

_____ 1. Which factor is the variable in this experiment?
   A the type of bee
   B the time of day
   C the type of plant
   D the color of the flowers

_____ 2. Which of the following hypotheses could be tested by this experiment?
   F Honeybees prefer to visit rose plants.
   G Honeybees prefer to visit red flowers.
   H Honeybees prefer to visit flowers in the morning.
   I Honey bees prefer to visit Portland rose flowers between 9 and 10 a.m.

_____ 3. Which of the following is the best reason why the Portland rose plant is included in all of the groups to be studied?
   A The type of plant is a control factor; any type of flowering plant could be used as long as all plants were of the same type.
   B The experiment will test whether bees prefer the Portland rose over other flowers.
   C An experiment should always have more than one variable.
   D The Portland rose is a very common plant.

MATH

Read each question below, and choose the best answer.

_____ 1. A survey of students was conducted to find out how many people were in each student's family. The replies from five students were as follows: 3, 3, 4, 4, and 6. What was the average family size?
   A 3
   B 3.5
   C 4
   D 5

_____ 2. In the survey above, if one more student were surveyed, which reply would make the average lower?
   F 3
   G 4
   H 5
   I 6

_____ 3. If an object that is 5 µm long were magnified by 1,000, how long would that object then appear?
   A 5 µm
   B 5 mm
   C 1,000 µm
   D 5,000 mm

_____ 4. How many meters are in 50 km?
   F 50 m
   G 500 m
   H 5,000 m
   I 50,000 m

_____ 5. What is the area of a square whose sides measure 4 m each?
   A 16 m
   B 16 m^2
   C 32 m
   D 32 m^2

Name _____ Class _____ Date _____

Skills Practice Lab

DATASHEET FOR CHAPTER LAB

Does It All Add Up?

Your math teacher won't tell you this, but did you know that sometimes 2 + 2 does not appear to equal 4?! In this experiment, you will use scientific methods to predict, measure, and observe the mixing of two unknown liquids. You will learn that a scientist does not set out to prove a hypothesis but to test it and that sometimes the results just don't seem to add up!

OBJECTIVES

Apply scientific methods to predict, measure, and observe the mixing of two unknown liquids.

MATERIALS
- beakers, 100 mL (2)
- Celsius thermometer
- glass-labeling marker
- graduated cylinders, 50 mL (3)
- liquid A, 75 mL
- liquid B, 75 mL
- protective gloves

SAFETY INFORMATION

MAKE OBSERVATIONS

1. Put on your safety goggles, gloves, and lab apron. Examine the beakers of liquids A and B provided by your teacher. Write down as many observations as you can about each liquid.

_____

_____

Caution: Do not taste, touch, or smell the liquids.

2. Pour exactly 25 mL of liquid A from the beaker into each of two 50 mL graduated cylinders. Combine these samples in one of the graduated cylinders. Record the final volume. Pour the liquid back into the beaker of liquid A. Rinse the graduated cylinders. Repeat this step for liquid B.

FORM A HYPOTHESIS

3. Based on your observations and on prior experience, formulate a testable hypothesis that states what you expect the volume to be when you combine 25 mL of liquid A with 25 mL of liquid B.

_____

_____

Name _____ Class _____ Date _____

Does It All Add Up? continued

4. Make a prediction based on your hypothesis. Use an if-then format. Explain why you made your prediction.

_____

_____

_____

Data Table

	Contents of cylinder A	Contents of cylinder B	Mixing results: predictions	Mixing results: observations
Volume				
Appearance				
Temperature				

TEST THE HYPOTHESIS

5. Use the table above to record your data.

6. Mark one graduated cylinder "A." Carefully pour exactly 25 mL of liquid A into this cylinder. In your data table, record its volume, appearance, and temperature.

7. Mark another graduated cylinder "B." Carefully pour exactly 25 mL of liquid B into this cylinder. Record its volume, appearance, and temperature in your data table.

8. Mark the empty third cylinder "A 1 B."

9. In the "Mixing results: predictions" column in your table, record the prediction you made earlier. Each classmate may have made a different prediction.

10. Carefully pour the contents of both cylinders into the third graduated cylinder.

11. Observe and record the total volume, appearance, and temperature in the "Mixing results: observations" column of your table.

ANALYZE THE RESULTS

1. **Analyzing Data** Discuss your predictions as a class. How many different predictions were there? Which predictions were supported by testing? Did any measurements surprise you?

_____

_____

_____

_____

Name _____ Class _____ Date _____

Does It All Add Up? *continued*

DRAW CONCLUSIONS

2. **Drawing Conclusions** Was your hypothesis supported or disproven? Either way, explain your thinking. Describe everything that you think you learned from this experiment.

_____

_____

3. **Analyzing Methods** Explain the value of incorrect predictions.

_____

_____

Name _____ Class _____ Date _____

Quick Lab

DATASHEET FOR QUICK LAB

Measure Up!

MATERIALS

- balance or scale, metric
- graduated cylinder, 25 or 100 mL
- markers, assorted colors
- meterstick
- posterboard
- ruler, metric
- thermometer, safety, celsius

SAFETY

1. For each of the following tasks, find a different item to measure. With permission from your teacher or parent, you may choose items within your classroom, school, or home.

   a. Measure length with a **meterstick**.
   b. Measure length with a **metric ruler**.
   c. Measure and calculate area in square meters.
   d. Measure volume with a **graduated cylinder**.
   e. Measure and calculate volume in cubic meters.
   f. Measure mass with a **balance**.
   g. Measure temperature with a **thermometer**.

   _____
   _____
   _____
   _____
   _____
   _____
   _____

2. Make a **poster** to present your measurements. Include drawings showing how you measured each item and tips stating how to use the measurement tools properly.

Name _____ Class _____ Date _____

Skills Practice Lab
Graphing Data

DATASHEET FOR LABBOOK

When performing an experiment, you usually need to collect data. To understand the data, you can often organize them into a graph. Graphs can show trends and patterns that you might not notice in a table or list. In this exercise, you will practice collecting data and organizing the data into a graph.

MATERIALS

- beaker, 400 mL
- clock (or watch) with a second hand
- gloves, heat-resistant
- hot plate
- ice
- paper, graph
- thermometer, Celsius, with a clip
- water, 200 mL

SAFETY INFORMATION

PROCEDURE

1. Pour 200 mL of water into a 400 mL beaker. Add ice to the beaker until the waterline is at the 400 mL mark.

2. Place a Celsius thermometer into the beaker. Use a thermometer clip to prevent the thermometer from touching the bottom of the beaker. Record the temperature of the ice water.

3. Place the beaker and thermometer on a hot plate. Turn the hot plate on medium heat, and record the temperature every minute until the water temperature reaches 100°C.

4. Using heat-resistant gloves, remove the beaker from the hot plate. Continue to record the temperature of the water each minute for 10 more minutes. Caution: Don't forget to turn off the hot plate.

5. On a piece of graph paper, create a graph similar to the one on the next page. Label the horizontal axis (the x-axis) "Time (min)," and mark the axis in increments of 1 min as shown. Label the vertical axis (the y-axis) "Temperature (°C)," and mark the axis in increments of 10° as shown.

Graphing Data continued

6. Find the 1 min mark on the x-axis, and move up the graph to the temperature you recorded at 1 min. Place a dot on the graph at that point. Plot each temperature in the same way. When you have plotted all of your data, connect the dots with a smooth line.

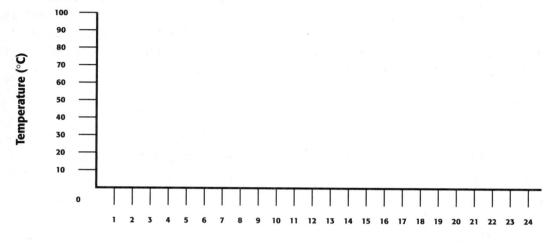

ANALYZE THE RESULTS

1. Examine your graph. Do you think the water heated faster than it cooled? Explain.

_____

_____

2. Estimate what the temperature of the water was 2.5 min after you placed the beaker on the hot plate. Explain how you can make a good estimate of temperature between those you recorded.

_____

_____

DRAW CONCLUSIONS

3. Explain how a graph may give more information than the same data in a table.

_____

_____

_____

_____

_____

Name _____ Class _____ Date _____

Model-Making Lab

A Window to a Hidden World

DATASHEET FOR LABBOOK

Have you ever noticed that objects underwater appear closer than they really are? The reason is that light waves change speed when they travel from air into water. Anton van Leeuwenhoek, a pioneer of microscopy in the late 17th century, used a drop of water to magnify objects. That drop of water brought a hidden world closer into view. How did Leeuwenhoek's microscope work? In this investigation, you will build a model of it to find out.

MATERIALS

- eyedropper
- hole punch
- newspaper
- plastic wrap, clear
- poster board, 3 cm × 10 cm
- tape, transparent
- water

PROCEDURE

1. Punch a hole in the center of the poster board with a hole punch.
2. Tape a small piece of clear plastic wrap over the hole. Be sure the plastic wrap is large enough so that the tape you use to secure it does not cover the hole.
3. Use an eyedropper to put one drop of water over the hole. Check to be sure your drop of water is dome shaped (convex).
4. Hold the microscope close to your eye and look through the drop. Be careful not to disturb the water drop.
5. Hold the microscope over a piece of newspaper, and observe the image.

ANALYZE THE RESULTS

1. Describe and draw the image you see. Is the image larger than or the same size as it is without the microscope? Is the image clear or blurred? Is the shape of the image distorted?

_____
_____
_____

A Window to a Hidden World continued

DRAW CONCLUSIONS

2. How do you think your model could be improved?

_____

_____

_____

APPLYING YOUR DATA

Robert Hooke and Zacharias Janssen contributed much to the field of microscopy. Research one of them, and write a paragraph about his contributions.

_____

_____

_____

_____

_____

_____

Name _____ Class _____ Date _____

Activity
Vocabulary Activity

The Puzzling World of Life Science

After you finish reading the chapter, try this puzzle! Using each of these clues, fill in the blanks provided on the next page with the letters of the word or phrase described below.

1. the use of knowledge, tools, and materials to solve problems and accomplish tasks
2. a unifying explanation for a broad range of hypotheses and observations that have been supported by testing
3. the measure of an object's surface
4. the amount of space that something occupies
5. a(n) _____ tests one factor at a time
6. the study of living things
7. a summary of many experimental results and observations
8. the one factor that differs in a controlled experiment
9. series of steps scientists use to answer a question or solve a problem
10. measured in kelvins or degrees celsius
11. the amount of matter that makes up an object
12. a(n) _____ has a tube with lenses, a stage, and a light source
13. possible answer to a question
14. uses electrons to produce magnified images
15. a representation of an object or a system

Name _____ Class _____ Date _____

Vocabulary Activity continued

1. _ _ _ _ _ _ _ _ _ _
    1
2. _ _ _ _ _ _
       2
3. _ _ _ _
       3
4. _ _ _ _ _ _
         4
5. _ _ _ _ _ _ _ _ _ _   _ _ _ _ _ _ _ _ _ _
                  5
6. _ _ _ _   _ _ _ _ _ _
       6
7. _ _ _
    7
8. _ _ _ _ _ _ _
         8
9. _ _ _ _ _ _ _ _ _   _ _ _ _ _ _ _
              9
10. _ _ _ _ _ _ _ _ _ _
    10
11. _ _ _ _
      11
12. _ _ _ _ _ _ _   _ _ _ _   _ _ _ _ _ _ _ _ _ _
    12a   12b
13. _ _ _ _ _ _ _ _ _
           13
14. _ _ _ _ _ _ _   _ _ _ _ _ _ _ _ _
           14a   14b
15. _ _ _ _ _
       15

Discover the phrase below by filling in the in the blanks with the letters above the matching numbers.

16. _ _ _   S T _ _ Y O _   _ _ _ _   _ _ _ _ _ _ _
     1 2 3   4 5   6 7 8 9 10 11 12a 12b 13 14a 14b 15

Activity
SciLinks Activity

SI UNITS

Go to www.scilinks.org. To find links related to the International System of Units, type in the keyword HSM1390. Then, use the links to fill in the following chart with information about the International System of Units.

Go to www.scilinks.org
Topic: SI Units
SciLinks code: HSM1390

Physical Quantity	Unit Name	Unit Symbol	Description

Performance-Based Assessment

Teacher Notes and Answer Key

PURPOSE

Students will use the scientific method to find out why pennies minted in different years have different masses.

Jennifer Ford
Northridge Middle School
North Richland Hills, Texas

TIME REQUIRED

One 45-minute class period. Students will need 35 minutes to perform the procedure and 10 minutes to answer the analysis questions.

RATING

Easy ←―1―――2―――3―――4―→ Hard

Teacher Prep–3
Student Set-Up–2
Concept Level–2
Clean Up–1

ADVANCE PREPARATION

Equip each student activity station with the necessary materials.

SAFETY INFORMATION

Have a disposal container for sharps nearby in case of glass breakage. Wipe up all spills immediately to avoid slipping.

TEACHING STRATEGIES

This activity works best in groups of 2–3 students. A quick lesson on how to read a meniscus may help students accurately complete the activity. Students should be familiar with the concept of density in order to successfully complete the activity.

Performance-Based Assessment *continued*

Evaluation Strategies

Use the following rubric to help evaluate student performance.

Rubric for Assessment

Possible points	Appropriate use of materials and equipment (20 points possible)
20–10	Safe and careful handling of materials and equipment; attention to detail; excellent lab skills
9–1	Little attention to detail; apparent lack of skill
	Ability to create hypotheses to explain mass differences (20 points possible)
20–10	Able to form a testable hypothesis; hypothesis stated unclearly; shows illogical thinking
9–1	Unable to form a testable hypothesis; hypothesis stated unclearly; shows illogical thinking
	Use of scientific methods (40 points possible)
40–30	Superior use of scientific method; each step is clearly understood and correctly implemented
29–20	Adequate use of scientific method; some steps are unclear or incorrect, minor difficulty in expression
19–1	Poor use of the scientific method; shows little effort to understand or participate
	Ability to record results (20 points possible)
20–10	Results are recorded neatly, completely, and accurately
9–1	Results are unclear, incomplete, or inaccurate

Name _____ Class _____ Date _____

Assessment

SKILLS PRACTICE

Performance-Based Assessment

OBJECTIVE

Using the scientific method, you will find out why pennies minted in different years have different masses.

KNOW THE SCORE!

As you work through the activity, keep in mind that you will be earning a grade for the following:

- how well you work with materials and equipment (20%)
- how well you formulate hypotheses (20%)
- how well you use the scientific method to prove or disprove your hypotheses (40%)
- how well you record the results of your experiments (20%)

Using Scientific Methods

ASK A QUESTION

Why are pennies from some years heavier than pennies from others?

MATERIALS AND EQUIPMENT

- 12 pennies: 4 from the 1970s, 4 from the 1980s, and 4 from the 1990s
- metric balance
- 50 mL graduated cylinder
- water

SAFETY INFORMATION

- Be careful with glass objects.
- Wipe up all spills immediately to avoid slipping.

FORM A HYPOTHESIS

1. Form a hypothesis to answer the question.

   Sample answer: Older pennies are heavier because they are made of denser metals.

Name _____ Class _____ Date _____

Performance-Based Assessment *continued*

TEST THE HYPOTHESIS

2. Arrange the pennies in chronological order, from oldest to newest. Starting with the oldest penny, find the mass of each penny and record each mass next to the year in the data chart provided.

3. Fill the graduated cylinder about half full with water. Record the exact volume in the first row of the data chart.

4. Place the first penny in the cylinder and record the new water level in the data chart.

5. Subtract the water level before the penny was added from the water level after the penny was added. This gives you the volume of the penny. Then calculate the density of the penny by dividing the mass by the volume. Record this value in the chart.

6. Repeat this procedure for each of the remaining pennies, using the water level from the previous step as the new starting level. Do not remove any pennies from the cylinder.

Data Collection Table

Year	Mass (g)	Water level before penny is added (mL)	Water level after penny (mL)	Difference in water levels	Density (g/mL)
1972	3.4	50.0	50.5	0.5	6.8
1974	3.3	50.5	51.0	0.5	6.6
1975	3.4	51.0	51.6	0.6	5.7
1977	3.6	51.6	52.1	0.5	7.2
1981	3.2	52.1	52.8	0.7	4.6
1982	3.5	52.8	53.5	0.7	5.0
1985	2.5	53.5	54.0	0.5	5.0
1986	2.5	54.0	54.9	0.9	2.7
1990	2.6	54.9	55.5	0.6	4.3
1992	2.3	55.5	56.0	0.5	4.6
1995	2.5	56.0	56.5	0.5	5.0
1998	2.8	56.5	57.1	0.6	4.7

ANALYZE THE RESULTS

7. What does this experiment tell you about the volume of pennies?

Sample answer: The volume of pennies has not changed significantly over the past 30 years.

Name _____ Class _____ Date _____

Performance-Based Assessment continued

8. How has the density of pennies changed over the past 30 years?

Newer pennies are less dense than older pennies

9. Why might the density have changed?

Sample answer: Pennies may be made of less dense metals than they used to be.

| Skills Practice Lab

Does It All Add Up?

DATASHEET FOR CHAPTER LAB

Teacher Notes and Answer Key

TIME REQUIRED One 45-minute class period

LAB RATINGS Easy ←— 1 2 3 4 —→ Hard
   Teacher Prep–2
   Student Set-Up–1
   Concept Level–2
   Clean Up–1

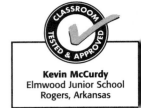

Kevin McCurdy
Elmwood Junior School
Rogers, Arkansas

MATERIALS

The materials listed on the student page are enough for a group of 2–3 students. Prepare a large jug of plain water labeled "Liquid A" and a jug of either isopropyl alcohol (2-propanol, $CH_3CH(OH)CH_3$) or denatured ethyl alcohol (95% ethanol, CH_3CH_2OH) labeled "Liquid B." Safety thermometers are recommended for this lab.

SAFETY CAUTION

- Remind students to review all safety cautions and icons before beginning this activity.

CAUTION

- Students should handle thermometers with care, and should treat all unknown chemicals as dangerous.
- Alcohol is flammable and poisonous. Students should wear goggles and aprons at all times.
- A fire extinguisher and fire blanket should be nearby. Know how to use them.
- The room should be well-ventilated, and students should be familiar with evacuation procedures.

LAB NOTES

Do not reveal the identity of the liquids until the end of the lab!
In this lab, students will likely be surprised to discover that 25 mL of liquid A (water) plus 25mL of liquid B (an alcohol) do not make 50 mL of the mixture. Spaces between molecules of alcohol become filled with water molecules, resulting in a lower total volume. The water-alcohol mixture will be cloudy and bubbly for a brief time after mixing, and may emit some heat. Have students record observations until the mixture becomes clear and then make their final measurements and observations.

TRY THE FOLLOWING DEMONSTRATION

In order to model the mixing of water and alcohol molecules for your students. Mix 25 mL of marbles with 25 mL of round BB-gun pellets. The BBs will settle between the marbles, and the result will be a total volume less than 50 mL.

Copyright © by Holt, Rinehart and Winston. All rights reserved.

Holt Science and Technology   The World of Life Science

Name _____ Class _____ Date _____

Skills Practice Lab

DATASHEET FOR CHAPTER LAB

Does It All Add Up?

Your math teacher won't tell you this, but did you know that sometimes 2 + 2 does not appear to equal 4?! In this experiment, you will use scientific methods to predict, measure, and observe the mixing of two unknown liquids. You will learn that a scientist does not set out to prove a hypothesis but to test it and that sometimes the results just don't seem to add up!

OBJECTIVES

Apply scientific methods to predict, measure, and observe the mixing of two unknown liquids.

MATERIALS

- beakers, 100 mL (2)
- Celsius thermometer
- glass-labeling marker
- graduated cylinders, 50 mL (3)
- liquid A, 75 mL
- liquid B, 75 mL
- protective gloves

SAFETY INFORMATION

MAKE OBSERVATIONS

1. Put on your safety goggles, gloves, and lab apron. Examine the beakers of liquids A and B provided by your teacher. Write down as many observations as you can about each liquid.

   All answers to this lab are based on student observations and may vary.

   _____

   Caution: Do not taste, touch, or smell the liquids.

2. Pour exactly 25 mL of liquid A from the beaker into each of two 50 mL graduated cylinders. Combine these samples in one of the graduated cylinders. Record the final volume. Pour the liquid back into the beaker of liquid A. Rinse the graduated cylinders. Repeat this step for liquid B.

FORM A HYPOTHESIS

3. Based on your observations and on prior experience, formulate a testable hypothesis that states what you expect the volume to be when you combine 25 mL of liquid A with 25 mL of liquid B.

   All answers to this lab are based on student observations and may vary.

   _____

Name _____ Class _____ Date _____

Does It All Add Up? continued

4. Make a prediction based on your hypothesis. Use an if-then format. Explain why you made your prediction.

   All answers to this lab are based on student observations and may vary.

   _____

   _____

Data Table

	Contents of cylinder A	Contents of cylinder B	Mixing results: predictions	Mixing results: observations
Volume				
Appearance				
Temperature				

TEST THE HYPOTHESIS

5. Use the table above to record your data.

6. Mark one graduated cylinder "A." Carefully pour exactly 25 mL of liquid A into this cylinder. In your data table, record its volume, appearance, and temperature.

7. Mark another graduated cylinder "B." Carefully pour exactly 25 mL of liquid B into this cylinder. Record its volume, appearance, and temperature in your data table.

8. Mark the empty third cylinder "A 1 B."

9. In the "Mixing results: predictions" column in your table, record the prediction you made earlier. Each classmate may have made a different prediction.

10. Carefully pour the contents of both cylinders into the third graduated cylinder.

11. Observe and record the total volume, appearance, and temperature in the "Mixing results: observations" column of your table.

ANALYZE THE RESULTS

1. **Analyzing Data** Discuss your predictions as a class. How many different predictions were there? Which predictions were supported by testing? Did any measurements surprise you?

   Students may make some unusual predictions. You may want to lead them

   into questions about volume. Encourage them to think of many ways to

   observe and characterize the two liquids. Avoid giving away the the

   explanation too quickly.

Name _____ Class _____ Date _____

Does It All Add Up? continued

DRAW CONCLUSIONS

2. Drawing Conclusions Was your hypothesis supported or disproven? Either way, explain your thinking. Describe everything that you think you learned from this experiment.

Check that students are clear about whether or not their hypothesis was correct and in what ways their observations supported or disproved their hypothesis.

3. Analyzing Methods Explain the value of incorrect predictions.

Incorrect predictions can lead to new questions and a new understanding of the way things work.

Name _____ Class _____ Date _____

Quick Lab

DATASHEET FOR QUICK LAB

Measure Up!

MATERIALS

- balance or scale, metric
- graduated cylinder, 25 or 100 mL
- markers, assorted colors
- meterstick
- posterboard
- ruler, metric
- thermometer, safety, celsius

SAFETY

1. For each of the following tasks, find a different item to measure. With permission from your teacher or parent, you may choose items within your classroom, school, or home.
    a. Measure length with a **meterstick**.
    b. Measure length with a **metric ruler**.
    c. Measure and calculate area in square meters.
    d. Measure volume with a **graduated cylinder**.
    e. Measure and calculate volume in cubic meters.
    f. Measure mass with a **balance**.
    g. Measure temperature with a **thermometer**.

    Check that students record their observations accurately.

    _____
    _____
    _____
    _____
    _____

2. Make a **poster** to present your measurements. Include drawings showing how you measured each item and tips stating how to use the measurement tools properly.

    Check that student posters portray correct procedures.

Safety Caution: Supervise and direct students carefully in the safe and responsible use of all equipment. Non-mercury safety thermometers and plastic graduated cylinders are recommended. Teacher Note: This activity presents an opportunity to see what knowledge and skills students already have related to using basic equip-ment to perform measurements. Take the opportunity to observe, guide, and correct the students' methods. Suggest challenging and interesting objects for the students to measure.

Copyright © by Holt, Rinehart and Winston. All rights reserved.

Holt Science and Technology

The World of Life Science

Skills Practice Lab

Graphing Data

DATASHEET FOR LABBOOK

Teacher Notes and Answer Key

TIME REQUIRED
One 45-minute class period

Edith C. McAlanis
Socorro Middle School
El Paso, Texas

LAB RATINGS

Easy ← 1   2   3   4 → Hard

Teacher Prep–2
Student Set-Up–1
Concept Level–1
Clean Up–1

MATERIALS
The materials listed on the student page are enough for a group of 4–5 students. A 3 × 5 in. index card cut in half lengthwise can substitute for a stiff piece of poster board. It can be difficult to eliminate wrinkles in the plastic over the hole. Some students may need assistance.

SAFETY INFORMATION

Safety Caution Caution students to exercise proper care when handling the beaker of hot water. Also, caution students to be careful when they are moving around an electrical cord. A clip that will hold the thermometer to the side of the beaker and off the bottom of the beaker while it is heating or cooling is safer and more accurate than a thermometer simply propped up inside the beaker.

Name _____ Class _____ Date _____

Skills Practice Lab

Graphing Data

DATASHEET FOR LABBOOK

When performing an experiment, you usually need to collect data. To understand the data, you can often organize them into a graph. Graphs can show trends and patterns that you might not notice in a table or list. In this exercise, you will practice collecting data and organizing the data into a graph.

MATERIALS

- beaker, 400 mL
- clock (or watch) with a second hand
- gloves, heat-resistant
- hot plate
- ice
- paper, graph
- thermometer, Celsius, with a clip
- water, 200 mL

SAFETY INFORMATION

PROCEDURE

1. Pour 200 mL of water into a 400 mL beaker. Add ice to the beaker until the waterline is at the 400 mL mark.

2. Place a Celsius thermometer into the beaker. Use a thermometer clip to prevent the thermometer from touching the bottom of the beaker. Record the temperature of the ice water.

3. Place the beaker and thermometer on a hot plate. Turn the hot plate on medium heat, and record the temperature every minute until the water temperature reaches 100°C.

4. Using heat-resistant gloves, remove the beaker from the hot plate. Continue to record the temperature of the water each minute for 10 more minutes. Caution: Don't forget to turn off the hot plate.

5. On a piece of graph paper, create a graph similar to the one on the next page. Label the horizontal axis (the x-axis) "Time (min)," and mark the axis in increments of 1 min as shown. Label the vertical axis (the y-axis) "Temperature (°C)," and mark the axis in increments of 10° as shown.

Graphing Data continued

6. Find the 1 min mark on the x-axis, and move up the graph to the temperature you recorded at 1 min. Place a dot on the graph at that point. Plot each temperature in the same way. When you have plotted all of your data, connect the dots with a smooth line.

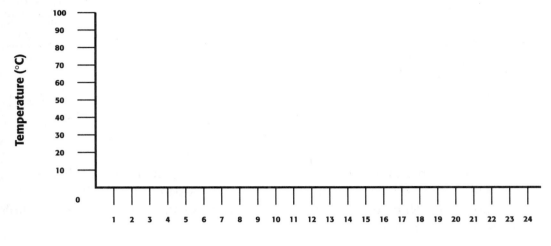

ANALYZE THE RESULTS

1. Examine your graph. Do you think the water heated faster than it cooled? Explain.

Answers may vary according to several factors, including altitude.

2. Estimate what the temperature of the water was 2.5 min after you placed the beaker on the hot plate. Explain how you can make a good estimate of temperature between those you recorded.

Answers may vary.

DRAW CONCLUSIONS

3. Explain how a graph may give more information than the same data in a table.

A list or a chart is organized information, and sometimes it is necessary to put collected data into one of these forms before graphing. Because a graph is like a picture, it can often help scientists to see what is happening when numbers alone would be confusing. A graph can show a trend or a pattern that may not be readily discernible in a list or chart.

Model-Making Lab

DATASHEET FOR LABBOOK

A Window to a Hidden World

Teacher Notes and Answer Key

TIME REQUIRED

One 45-minute class period

LAB RATINGS

Easy ←—1——2——3——4—→ Hard

    Teacher Prep–2
    Student Set-Up–2
    Concept Level–1
    Clean Up–1

Georgiann Delgadillo
East Valley School District
Continuous Curriculum School
Spokane, Washington

MATERIALS

The materials listed on the student page are enough for a group of 4–5 students. A 3 × 5 in. index card cut in half lengthwise can substitute for a stiff piece of poster board. It can be difficult to eliminate wrinkles in the plastic over the hole. Some students may need assistance.

Name _____ Class _____ Date _____

Model-Making Lab

DATASHEET FOR LABBOOK

A Window to a Hidden World

Have you ever noticed that objects underwater appear closer than they really are? The reason is that light waves change speed when they travel from air into water. Anton van Leeuwenhoek, a pioneer of microscopy in the late 17th century, used a drop of water to magnify objects. That drop of water brought a hidden world closer into view. How did Leeuwenhoek's microscope work? In this investigation, you will build a model of it to find out.

MATERIALS

- eyedropper
- hole punch
- newspaper
- plastic wrap, clear
- poster board, 3 cm × 10 cm
- tape, transparent
- water

PROCEDURE

1. Punch a hole in the center of the poster board with a hole punch.
2. Tape a small piece of clear plastic wrap over the hole. Be sure the plastic wrap is large enough so that the tape you use to secure it does not cover the hole.
3. Use an eyedropper to put one drop of water over the hole. Check to be sure your drop of water is dome shaped (convex).
4. Hold the microscope close to your eye and look through the drop. Be careful not to disturb the water drop.
5. Hold the microscope over a piece of newspaper, and observe the image.

ANALYZE THE RESULTS

1. Describe and draw the image you see. Is the image larger than or the same size as it is without the microscope? Is the image clear or blurred? Is the shape of the image distorted?

   Answers will vary. Students should see a slightly larger image. It will be

   blurred, especially around the edges. The image may be distorted.

Name _____ Class _____ Date _____

A Window to a Hidden World continued

DRAW CONCLUSIONS

2. How do you think your model could be improved?

Some students may think their model could be improved by eliminating the wrinkles over the hole.

APPLYING YOUR DATA

Robert Hooke and Zacharias Janssen contributed much to the field of microscopy. Research one of them, and write a paragraph about his contributions.

Robert Hooke (1635–1703), one of the world's great inventors, is famous for his discovery of "cells" in cork tissue as seen through his improved microscope. Hooke was also a keen observer with an interest in fossils and geology. Zacharias Janssen, a Dutch lens grinder, mounted two lenses in a tube to produce the first compound microscope in 1590.

Answer Key

Directed Reading A

SECTION: ASKING ABOUT LIFE

1. life science
2. diversity
3. Answers will vary. Sample answer: Where does it live?
4. anyone
5. anywhere—in a laboratory, on farms, in forests, on the ocean floor, in space, for business, hospitals, government agencies, universities, as teachers
6. his or her curiosity
7. AIDS
8. inherited
9. Many environmental problems are caused by people's misuse of natural resources.
10. By learning about the needs of Siberian tigers, scientists hope to develop a plan that will help them survive.

SECTION: SCIENTIFIC METHODS

1. making observations; forming a hypothesis
2. The order depends on what works best to answer their questions.
3. B
4. hypothesis
5. testable
6. prediction
7. an experiment that tests only one factor at a time by using a comparison of a control group with one or more experimental groups
8. a factor that differs between the control group and experimental groups
9. planning
10. The more organisms tested, the more certain scientists can be of the data they collect in an experiment.
11. Answers will vary. Sample answer: in a graph or a data table
12. Both are helpful. Either way, the scientist has learned something, which is the purpose of using scientific methods.
13. another investigation
14. Answers will vary. Sample answer: so other scientists may repeat the experiments to see if they get the same results; so scientists with similar interests can compare hypotheses and form consistent explanations

SECTION: SCIENTIFIC MODELS

1. C
2. C
3. B
4. A
5. D
6. theory
7. law
8. scientific theories are used to explain observations and to predict what might happen: Scientific law is a summary of observations and tells us what will happen.
9. theories, facts, laws

SECTION: TOOLS, MEASUREMENT, AND SAFETY

1. Life scientists use tools to make observations and to gather, store, and analyze information.
2. the application of science for practical purposes
3. 1946
4. to create graphs, solve complex equations, analyze data, share data and ideas, and publish reports
5. B
6. D
7. C
8. A
9. C
10. It helps scientists share and compare observations and results. Almost all units are based on the number 10, which makes conversions easy.
11. millimeter
12. area
13. square
14. volume
15. liters, milliliters, cubic meters, cubic centimeters, and cubic millimeters
16. mass

17. kilograms, metric tons, and grams
18. temperature
19. degrees Celsius, degrees Fahrenheit, and kelvins
20. Sample answer: Ask teacher for permission and read the lab procedure carefully.
21. B
22. D
23. C
24. G
25. F
26. I
27. E
28. A
29. H

Directed Reading B

SECTION: ASKING ABOUT LIFE

1. B
2. A
3. D
4. D
5. C
6. B
7. diseases
8. inherited
9. tigers

SECTION: SCIENTIFIC METHODS

1. scientific methods
2. asking questions
3. counting
4. accurate
5. hypothesis
6. if-then
7. factor
8. controlled experiment
9. variable
10. A
11. A
12. A
13. C
14. A
15. B

SECTION: SCIENTIFIC MODELS

1. A
2. B
3. A
4. C
5. A
6. A
7. B
8. B
9. B
10. D
11. A

SECTION: TOOLS, MEASUREMENT, AND SAFETY

1. tools
2. technology
3. computer
4. analyze data
5. B
6. D
7. C
8. A
9. B
10. C
11. A
12. A
13. C
14. D
15. A
16. B
17. D
18. B
19. C
20. A

Vocabulary and Section Summary

SECTION: ASKING ABOUT LIFE

1. life science: the study of living things

SECTION: SCIENTIFIC METHODS

1. scientific methods – a series of steps followed to solve problems
2. hypothesis: an explanation that is based on prior scientific research or observations and that can be tested
3. controlled experiment: an experiment that tests only one factor at a time by using a comparison of a control group with an experimental group
4. variable: a factor that changes in an experiment in order to test a hypothesis

SECTION: SCIENTIFIC MODELS

1. model: a pattern, plan, representation, or description designed to show the structure or workings of an object, system, or concept
2. theory: an explanation that ties together many hypotheses and observations

3. law: a summary of many experimental results and observations; a law tells how things work

SECTION: TOOLS, MEASUREMENT, AND SAFETY

1. technology: the application of science for practical purposes; the use of tools, machines, materials, and processes to meet human needs
2. compound light microscope: an instrument that magnifies small objects so that they can be seen easily by using two or more lenses
3. electron microscope: a microscope that focuses a beam of electrons to magnify objects
4. area: a measure of the size of a surface or a region
5. volume: a measure of the size of a body or region in three-dimensional space
6. mass: a measure of the amount of matter in an object
7. temperature: a measure of how hot (or cold) something is

Section Review

SECTION: ASKING ABOUT LIFE

1. Life science is the study of living things.
2. C
3. Scientific investigations usually start with a question.
4. Anywhere, and anything having to do with living things.
5. 15/50 = 30%
6. Sample answer: Eat breakfast —how digestion works; play with the dog —how animals behave; drive to school —environmental problems; play soccer —how muscles work, play chess—how the brain works.
7. Sample answer: How does the scientist get close to the tiger? Where do Siberian tigers live? What do Siberian tigers eat? Do Siberian tigers get cold? Are Siberian tigers the same as other tigers?

SECTION: SCIENTIFIC METHODS

1. Sample answer: A good controlled experiment will test a single hypothesis and a single variable at a time.
2. C
3. a control group and one or more experimental groups that differ by only one factor—the variable.
4. Scientific knowledge changes because scientists conduct new experiments to test new hypotheses, then share their results and build upon existing knowledge.
5. $(4 + 5 + 6 + 6 + 9) \div 5 = 30 \div 5 = 6$
6. Sample answer: Because the scientists were trying to test the hypothesis that UV light causes deformities; thus UV light was the factor that needed to be varied —the variable
7. Sample answer: Because sometimes scientists need to go back and change a step or they may be able to skip a step if someone else has already done it.
8. Sample answer: In addition to the variable being tested; other factors for the control group and the experimental group were not the same.
9. Sample answer: Observations are limited to our human senses or to what we have the technology to observe.
10. See sample graph below. The temperature that allows the bacteria to multiply most quickly is 37°C.

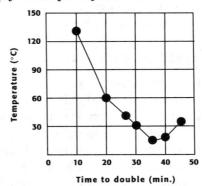

SECTION: SCIENTIFIC MODELS

1. A theory is an explanation that matches many hypotheses but may still change.
2. A law tells you exactly what to expect in certain situations.
3. B
4. Sample answer: Physical model: a plastic human body; Mathematical model: a Punnett square; Conceptual model: cell theory

5. Sample answer: Scientists use laws to predict what will happen under specific conditions but not to explain why scientists use theories to make general predictions and explain why they think something happens.
6. 0.21 m or 21 cm
7. Sample answer: The model could be used to determine if a certain prehistoric animal was tall enough to reach the leaves of the plant, or to show what the environment of a certain prehistoric area looked like. A limitation is that the model might not smell or taste like the real plant did.

SECTION: TOOLS, MEASUREMENT, AND SAFETY

1. area
2. mass
3. volume
4. B
5. Temperature is an indication of the amount of energy within matter.
6. grams or milligrams
7. 3.0 L = 3000 mL = 3000 cm^3
8. (28.5 x 22 x 3.5) = 2194.5 cm^3, or about 2,195 cm^3
9. An SEM. You can tell by the magnification and the 3-D features visible.
10. Sample answer: You could get hurt by an unknown chemical.

Chapter Review

1. Sample answer: You can use scientific methods to study life science.
2. Sample answer: A variable is a part of a controlled experiment.
3. Sample answer: A theory is an explanation for a broad range of observations and hypotheses. A hypothesis is an explanation of a specific set of observations and can be tested.
4. Sample answer: A compound light microscope uses light to create an image of an object, while an electron microscope uses electrons.
5. Sample answer: Area is a measure of a surface, while volume is a measure of three-dimensional size.
6. C
7. A
8. B
9. B
10. B
11. D
12. Sample answer: Science can be used to find cures for diseases, to understand animal behavior, and to solve environmental problems.
13. Hypotheses need to be testable in order to be useful; if no information can be gathered to either support or disprove a hypothesis, then it is merely an idea that cannot be built upon scientifically.
14. Sample answer: A life scientist studying animals might use a radio collar to track the animal's location, a computer database program to record data, and a computer mapping program to draw maps.
15. Sample answer: Physical models include toys and a model of a cell or human body. Mathematical models include a Punnett square and equations to calculate physical forces. Conceptual models include theories about how the solar system formed or how life evolved on Earth.
16. Sample answer: Advantages of models: they are easier to see and manipulate than the real thing might be, they can simplify concepts. Limitations: models do not behave exactly like the real thing, so they may not accurately predict results.
17. The SI units used to describe volume are liters (L), units based on the liter, and units based on the cubic meter (m^3, cm^3, mm^3). The SI units used to describe the mass of an object are kilograms (kg) and other units based on the gram.
18. The more individuals there are in the groups, the more confident scientists can be that differences between the groups were caused by the variable and not by natural differences between individual organisms.
19. An answer to this exercise can be found at the end of the book.
20. Sample answer: Observations are limited by human senses, or by technology. Things may exist that cannot be observed.

21. Sample answer: Vitamins help the mice remember the maze.
22. D
23. A
    (200mL − 125 mL = 75 mL = 75 cm^3)
24.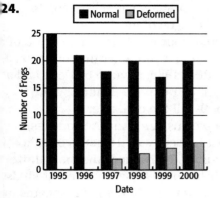

Reinforcement
THE MYSTERY OF THE BUBBLING TOP
6. Answers will vary. Sample answer: The plastic disk began to move on top of the bottle's neck.
7. Accept all reasonable responses. Sample starred answer: I think the plastic disk moved because the hot water warmed the air inside the bottle.
8. Answers will vary. Sample answer: I could try placing the bottle in a beaker of cold water to see if the plastic disk moves.
9. Answers will vary. Sample answer: The disk did not move.
10. Answers will vary. Sample answer: I think that the disk does not move because the cold water cannot heat up the air inside the bottle.
11. Sample answer: Yes; if the disk had started moving as soon as it did before, I would have known that the temperature of the water had nothing to do with the movement of the disk.
12. Sample answer: Yes; there are more factors that I could change. Next, I could change the temperature of the water inside the bottle to see what happens.

Critical Thinking
1. Citizens may get botulism by eating contaminated food from cans.
2. Answers will vary. Sample answer: The Health Department could provide a record of the number of bulging cans, the dates when the problem occurred, and whether *Clostridium botulinum* has been found in any cans.
3. Answers will vary. Sample answer: The bulging cans are a result of local canneries using a new canning process that allows the growth of bacteria.
4. Answers will vary. Sample answer: Collect two groups of cans from the local canneries. The groups will be identical except that only one group of cans will be processed using the new method. Observe the cans at the end the summer. Count the number of cans that have bulged in each group. Analyze the data, and determine whether the cans processed using the new method were significantly more likely to bulge.
5. Answers will vary. Answers should show a knowledge of the scientific methods and of the case. The proposed results should be consistent with the experiment and the original hypothesis.

Section Quizzes
SECTION: ASKING ABOUT LIFE
1. B   4. C
2. B   5. B
3. D   6. D

SECTION: SCIENTIFIC METHODS
1. B   4. C
2. A   5. D
3. A

SECTION: SCIENTIFIC MODELS
1. B   6. C
2. A   7. B
3. C   8. A
4. C   9. C
5. A   10. D

SECTION: TOOLS, MEASUREMENT, AND SAFETY
1. E   6. D
2. B   7. B
3. A   8. B
4. D   9. B
5. C   10. A

Chapter Test A

1. I	13. D
2. B	14. C
3. J	15. C
4. G	16. B
5. A	17. F
6. E	18. A
7. D	19. B
8. C	20. D
9. F	21. E
10. H	22. C
11. B	23. A
12. C	24. D

Chapter Test B

1. hypothesis
2. volume
3. compound light microscope
4. theory
5. technology
6. scientific methods
7. B
8. C
9. D
10. C
11. B
12. Sample answer: When a hypothesis is rejected, scientists learn that the hypothesis is not the correct explanation for the problem they studied. They may then seek a different explanation, reexamine data, or ask new questions.
13. Sample answer: The purpose of the control group is to isolate one factor in an experiment so that the factor's effect can be seen. The factor being tested is left unchanged in the control group, but is changed in the variable group. This factor can influence the outcome of the experiment.
14. Sample answer: The advantages include: they help explain how something works or describe how something is structured; they can represent things that are too small or too large to work with; and they can represent things that no longer exist. The disadvantage is that they are never exactly like the real thing.
15. Sample answer: If classical music improves a person's ability to study, then on average, people who listen to such music will score higher on a given test than people who study in silence or study while listening to another type of music.
16. Sample answer: Pour exactly 1 mL of a liquid into a container that has a length of 1 cm, a width of 1 cm, and a height of 1 cm. The liquid should exactly fill the container.
17. Sample answer: Plant A produces the most flowers with 16 hours of light and Plant B produces the most flowers with 10 hours of light. Different kinds of plants need different day lengths in order to produce flowers.
18. a. observations; b. hypothesis; c. technology; d. new questions; e. controlled experiments

Chapter Test C

1. B
2. B
3. B
4. C
5. A
6. C
7. C
8. A
9. D
10. B
11. A
12. C
13. D
14. B
15. A
16. C
17. D
18. model
19. scientific methods
20. controlled experiment
21. life science

Standardized Test Preparation

READING

Passage 1
1. C
2. F
3. C

Passage 2
1. B
2. I

INTERPRETING GRAPHICS
1. D
2. G
3. A

MATH
1. C
2. F
3. B
4. I
5. B

Vocabulary Activity
1. technology
2. theory
3. area
4. volume
5. controlled experiment
6. life science
7. law
8. variable
9. scientific methods
10. temperature
11. mass
12. compound light microscope
13. hypothesis
14. electron microscope
15. model
16. phrase: the study of life science

SciLinks Activity

Answers will vary. Students should list 7 symbols, providing information on the quantities being measured as well as the name of the unit, its symbol, and a description.

TEACHER RESOURCE PAGE

Lesson Plan

Section: Asking About Life

Pacing

Regular Schedule: with Lab(s): N/A   without Lab(s): 1 day

Block Schedule: with Lab(s): N/A   without Lab(s): 0.5 day

Objectives

1. Explain the importance of asking questions in life science.
2. State examples of life scientists at work.
3. Give three ways life science is beneficial to living things.

National Science Education Standards Covered

UCP 1: Systems, order, and organization

ST 2: Understandings about science and technology

SPSP 1: Personal health

SPSP 2: Populations, resources, and environments

SPSP 4: Risks and benefits

SPSP 5: Science and technology in society

HNS 1: Science as a human endeavor

HNS 2: Nature of science

HNS 3: History of science

KEY
SE = Student Edition   **TE** = Teacher's Edition
CRF = Chapter Resource File

FOCUS (5 minutes)

- **Chapter Starter Transparency** Use this transparency to introduce the chapter.
- **Bellringer, TE** Have students write five questions about the natural world and share their questions with the class.
- **Bellringer Transparency** Use this transparency as students enter the classroom and find their seats.

MOTIVATE (10 minutes)

- **Group Activity, Local Life Scientists, TE** Have groups brainstorm about what types of people in their community might be using life science and interview one of them. (**GENERAL**)

TEACHER RESOURCE PAGE

Lesson Plan

TEACH (20 minutes)

- **Reading Strategy, Paired Summarizing, SE** Have students utilize paired summarizing when reading the section.
- **Discussion, Why Ask Why? TE** Have students write about the benefits of studying how the human body responds to space travel and record their reasons. (**BASIC**)
- **Group Activity, Wildlife Safari, TE** Arrange a visit to a local zoo or wildlife area and have students ask and answer questions. (**GENERAL**)
- **Connection to Real World, Crime-Fighting Bugs, TE** Share with students the story of a forensic entomologist. (**GENERAL**)
- **Cultural Awareness, Historic Disease Researcher, TE** Tell students about Shibasaburo Kitasato's search for ways to fight disease. (**GENERAL**)
- **Inclusion Strategies, TE** Ask students to research questions about polio and other diseases.
- **Environmental Science Connection Activity, Pollution Demonstration, TE** Do a simple demonstration about pollution. (**BASIC**)
- **Directed Reading A/B, CRF** These worksheets reinforce basic concepts and vocabulary presented in the lesson. (**BASIC/SPECIAL NEEDS**)
- **Vocabulary and Section Summary, CRF** Students write definitions of key terms and read a summary of section content. (**GENERAL**)

CLOSE (10 minutes)

- **Reteaching, Asking Questions, TE** Display a picture and have students write questions about it. (**BASIC**)
- **Quiz, TE** Students answer 2 questions about life science. (**GENERAL**)
- **Alternative Assessment, Habitat Helpers, TE** Students investigate ways to preserve local habitats. (**GENERAL**)
- **Homework, Critter Comics, TE** Have students create an imaginary organism and environment and use them in a comic. (**GENERAL**)
- **Section Review, CRF** Students answer end-of-section vocabulary, key ideas, math, and critical thinking questions. (**GENERAL**)
- **Section Quiz, CRF** Students answer 6 objective questions about life science. (**GENERAL**)

TEACHER RESOURCE PAGE

Lesson Plan

Section: Scientific Methods

Pacing

Regular Schedule: with Lab(s): 2 days without Lab(s): 1 day

Block Schedule: with Lab(s): 1 day without Lab(s): 0.5 day

Objectives

1. Describe scientific methods.
2. Determine the appropriate design of a controlled experiment.
3. Use information in tables and graphs to analyze experimental results.
4. Explain how scientific knowledge can change.

National Science Education Standards Covered

UCP 2: Evidence, models, and explanation

UCP 3: Change, constancy, and measurement

SAI 1: Abilities necessary to do scientific inquiry

SAI 2: Understandings about scientific inquiry

SPSP2: Populations, resources, and environments

SPSP 5: Science and technology in society

HNS 1: Science as a human endeavor

HNS 2: Nature of science

HNS 3: History of science

KEY
SE = Student Edition   TE = Teacher's Edition
CRF = Chapter Resource File

FOCUS *(5 minutes)*

- **Bellringer, TE** Have students write an answer to a question and then discuss the question.
- **Bellringer Transparency** Use this transparency as students enter the classroom and find their seats.

MOTIVATE *(10 minutes)*

- **Activity, Now You See It, TE** Students practice making and recording observations after viewing shapes on an overhead projector. **(GENERAL)**

TEACHER RESOURCE PAGE

Lesson Plan

TEACH *(65 minutes)*

- **Reading Strategy, Reading Organizer, SE** Have students map a flowchart of possible steps in scientific methods.
- **Teaching Transparency, Scientific Methods** Use this graphic to introduce or review the steps of the scientific method.
- **Discussion, Testing Hypotheses, TE** Give students an observation, then have them write a hypothesis and propose a way to test it. **(BASIC)**
- **Activity, Test a Hypothesis, TE** Challenge students to design and carry out an investigation based on a hypothesis of their own. **(ADVANCED)**
- **Activity, Writing Predictions, TE** Have students practice writing predictions. **(GENERAL)**
- **Using the Table, Experimental Factors, TE** Provide extra help in reading the table about experimental factors. **(BASIC)**
- **Using the Figure, Control Group, TE** Help students understand the experimental setup shown in Figure 7. **(BASIC)**
- **Directed Reading A/B, CRF** These worksheets reinforce basic concepts and vocabulary presented in the lesson. **(BASIC/SPECIAL NEEDS)**
- **Reinforcement, CRF** This worksheet reinforces key concepts in the chapter by having students practice using scientific methods. **(GENERAL)**
- **Critical Thinking, CRF** This worksheet extends key concepts in the chapter by having students apply scientific methods to food safety. **(ADVANCED)**
- **Vocabulary and Section Summary, CRF** Students write definitions of key terms and read a summary of section content. **(GENERAL)**
- **Chapter Lab, Does It All Add Up?, SE** Students use scientific methods to predict, measure and observe the mixing of two unknown liquids. **(GENERAL)**
- **Datasheet for Chapter Lab, Does It All Add Up?, CRF** Have students use datasheet to complete the chapter lab. **(GENERAL)**

CLOSE *(10 minutes)*

- **Reteaching, Experimental Set-Up, TE** Have students propose other experiments scientists could do with the frogs and discuss them. **(BASIC)**
- **Quiz, TE** Students answer 2 questions about scientific methods. **(GENERAL)**
- **Alternative Assessment, Using Scientific Methods, TE** Students use scientific methods to answer a simple question. **(GENERAL)**
- **Homework, Investigate Your Area, TE** Have students make and record observations of a natural area. **(GENERAL)**
- **Section Review, CRF** Students answer end-of-section vocabulary, key ideas, math, and critical thinking questions. **(GENERAL)**
- **Section Quiz, CRF** Students answer 6 objective questions about scientific methods. **(GENERAL)**

Copyright © by Holt, Rinehart and Winston. All rights reserved.

TEACHER RESOURCE PAGE

Lesson Plan

Section: Scientific Models

Pacing

Regular Schedule: with Lab(s): N/A without Lab(s): 1 day

Block Schedule: with Lab(s): N/A without Lab(s): 0.5 day

Objectives

1. Give examples of three types of models.
2. Identify the benefits and limitations of models.
3. Compare the ways that scientists use hypotheses, theories, and laws.

National Science Education Standards Covered

UCP 2: Evidence, models, and explanation

UCP 3: Change, constancy, and measurement

SAI 1: Abilities necessary to do scientific inquiry

SAI 2: Understandings about scientific inquiry

SPSP 5: Science and technology in society

ST 2: Understandings about science and technology

HNS 1: Science as a human endeavor

HNS 2: Nature of science

HNS 3: History of science

KEY
SE = Student Edition   **TE** = Teacher's Edition
CRF = Chapter Resource File

FOCUS *(5 minutes)*

- **Bellringer, TE** Have students write answers to several questions and discuss their answers.

- **Bellringer Transparency** Use this transparency as students enter the classroom and find their seats.

MOTIVATE *(10 minutes)*

- **Discussion, Toys as Models, TE** Pass around various toys and have students discuss how toys serve as models. (**GENERAL**)

Copyright © by Holt, Rinehart and Winston. All rights reserved.

Holt Science and Technology   The World of Life Science

TEACHER RESOURCE PAGE

Lesson Plan

TEACH *(20 minutes)*

- **Reading Strategy, Reading Organizer, SE** Have students create an outline as they read the section.
- **Reading Strategy, Paired Reading TE** Have students read the section in pairs. (**BASIC**)
- **Group Activity, Classifying, TE** Have students create a classification system and illustrate it. (**GENERAL**)
- **Directed Reading A/B, CRF** These worksheets reinforce basic concepts and vocabulary presented in the lesson. (**BASIC/SPECIAL NEEDS**)
- **Vocabulary and Section Summary, CRF** Students write definitions of key terms and read a summary of section content. (**GENERAL**)

CLOSE *(10 minutes)*

- **Reteaching, Models, TE** Have students brainstorm a list of models and identify what type each is. (**BASIC**)
- **Quiz, TE** Students answer 2 questions about scientific models. (**GENERAL**)
- **Alternative Assessment, Tour Guide Talk, TE** Students write a speech for a guide who gives tours of a giant model cell. (**GENERAL**)
- **Homework, Computer-Generated Models, TE** Invite students to search the Internet for computer-generated models of prehistoric organisms. (**ADVANCED**)
- **Section Review, CRF** Students answer end-of-section vocabulary, key ideas, math, and critical thinking questions. (**GENERAL**)
- **Section Quiz, CRF** Students answer 10 objective questions about scientific models. (**GENERAL**)

TEACHER RESOURCE PAGE

Lesson Plan

Section: Tools, Measurement, and Safety

Pacing

Regular Schedule: with Lab(s): N/A without Lab(s): 1 day

Block Schedule: with Lab(s): N/A without Lab(s): 0.5 day

Objectives

1. Give three examples of how life scientists use computers and technology.

2. Describe three tools life scientists use to observe organisms.

3. Explain the importance of the International System of Units, and give four examples of SI units.

National Science Education Standards Covered

UCP 3: Change, constancy, and measurement

SAI 1: Abilities necessary to do scientific inquiry

SAI 2: Understandings about scientific inquiry

ST 1: Abilities of technological design

ST 2: Understandings about science and technology

KEY
SE = Student Edition **TE** = Teacher's Edition
CRF = Chapter Resource File

FOCUS *(5 minutes)*

- **Bellringer, TE** Have students write an answer to a question and then discuss their responses.
- **Bellringer Transparency** Use this transparency as students enter the classroom and find their seats.

MOTIVATE *(10 minutes)*

- **Demonstration, Tools For Seeing, TE** Display images produced with various types of technology and challenge students to identify them. **(GENERAL)**

TEACH *(20 minutes)*

- **Reading Strategy, Reading Organizer, SE** Have students make a concept map of the vocabulary terms as they read the section.
- **Activity, Using a Microscope, TE** Give students the opportunity to become more familiar with a light microscope by seeing its parts and viewing a slide. **(GENERAL)**

TEACHER RESOURCE PAGE

Lesson Plan

- **Group Activity, X Rays, TE** Have students imagine they were Roentgen's assistants and brainstorm a list of ways to use X rays. **(GENERAL)**
- **Teaching Transparency, Compound Light Microscope** Use this graphic to teach about the compound light microscope.
- **Cultural Awareness, It All Adds Up! TE** Have students prepare presentations about early calculating devices from other cultures. **(GENERAL)**
- **Debate, SI in the U.S., TE** Give students the opportunity to debate whether or not SI should be the only measurement system used in the U.S. **(GENERAL)**
- **Connection to Math, International System of Units, TE** Have students practice using the SI system. **(GENERAL)**
- **Teaching Transparency, Common SI Units and Conversions** Use this graphic to review SI units and conversions.
- **Demonstration, Measuring Mass and Volume, TE** Demonstrate to students the use and care of a graduated cylinder. **(BASIC)**
- **Inclusion Strategies, TE** Demonstrate the meaning of the term *irregular shape* through the use of objects of varying shapes. **(BASIC)**
- **Quick Lab, SE** Students measure or calculate length, area, volume, mass and temperature. **(GENERAL)**
- **Datasheet for Quick Lab, CRF** Have students use datasheet to complete the Quick Lab. **(GENERAL)**
- **Connection to Physical Science, How Hot, TE** Use the **Three Temperature Scales** teaching transparency to compare the various units for temperature. **(GENERAL)**
- **Directed Reading A/B, CRF** These worksheets reinforce basic concepts and vocabulary presented in the lesson. **(BASIC/SPECIAL NEEDS)**
- **Vocabulary and Section Summary, CRF** Students write definitions of key terms and read a summary of section content. **(GENERAL)**
- **SciLinks Activity, SI Units, SciLinks code HSM1390** Students research Internet resources related to the International System of Units. **(GENERAL)**

CLOSE *(10 minutes)*

- **Reteaching, SI Estimation, TE** Have students estimate the measurement of various objects in SI units. **(BASIC)**
- **Quiz, TE** Students answer 4 questions about tools, measurement, and safety. **(GENERAL)**
- **Alternative Assessment, Draw It, TE** Students create an illustrated dictionary of vocabulary terms. **(BASIC)**
- **Section Review, CRF** Students answer end-of-section vocabulary, key ideas, math, and critical thinking questions. **(GENERAL)**
- **Section Quiz, CRF** Students answer 10 objective questions about tools, measurement, and safety. **(GENERAL)**

TEACHER RESOURCE PAGE

Lesson Plan

End of Chapter Review and Assessment

Pacing

Regular Schedule: with Lab(s): N/A without Lab(s): 2 days

Block Schedule: with Lab(s): N/A without Lab(s): 1 day

KEY
SE = Student Edition TE = Teacher's Edition
CRF = Chapter Resource File

- **Chapter Review, CRF** Students answer end-of-chapter vocabulary, key ideas, critical thinking, and graphics questions. (**GENERAL**)
- **Vocabulary Activity, CRF** Students review chapter vocabulary terms by completing a puzzle. (**GENERAL**)
- **Concept Mapping Transparency, CRF** Students answer reading comprehension, math, and interpreting graphics questions in the formal of a standardized test.
- **Chapter Test A/B/C, CRF** Assign questions from the appropriate test for chapter assessment. (**GENERAL/ADVANCED/SPECIAL NEEDS**)
- **Performance-Based Assessment, CRF** Assign this activity for general level assessment for the chapter. (**GENERAL**)
- **Standardized Test Preparation, CRF** Students answer reading comprehension, math, and interpreting graphics questions in the format of a standardized test. (**GENERAL**)
- **Test Generator, One-Stop Planner** Create a customized homework assignment, quiz, or test using the HRW Test Generator Program.
- **CNN Video, CNN Presents Science in the News: Multicultural Connections,** Segment 1, Hopi Science
- **CNN Video, CNN Presents Science in the News: Scientists in Action,** Segment 1, A Biologist's Dolphin Investigation

TEST ITEM LISTING
The World of Life Science

MULTIPLE CHOICE

1. If you make an observation of a living thing and then ask a question about what you observed, you are
   a. noticing the diversity of life.
   b. behaving like a life scientist.
   c. solving a problem.
   d. learning how to protect the environment.
   Answer: B Difficulty: 1 Section: 1 Objective: 1

2. For every organism that has ever lived,
   a. there is only one question to ask.
   b. many questions could be asked.
   c. every question has already been asked.
   d. every question has already been answered.
   Answer: B Difficulty: 1 Section: 1 Objective: 1

3. Which of the following people is LEAST suited to being a life scientist?
   a. someone who likes sports.
   b. someone who goes to school.
   c. someone who is very curious.
   d. someone who has no interest in organisms.
   Answer: D Difficulty: 1 Section: 1 Objective: 2

4. A life scientist is LEAST likely to be found working
   a. in a laboratory.
   b. in a hospital.
   c. in an art museum.
   d. at the bottom of the ocean.
   Answer: C Difficulty: 1 Section: 1 Objective: 2

5. Which of the following is NOT one of the ways in which the work of a life scientist is beneficial?
   a. helping to fight diseases
   b. finding out about weather patterns
   c. studying environmental problems on Earth
   d. studying how humans inherit the code that controls their cells
   Answer: B Difficulty: 1 Section: 1 Objective: 3

6. Questions that life scientists today are trying to answer include all of the following EXCEPT
   a. the part of a person's inherited information responsible for certain inherited diseases.
   b. how the human body responds to space travel.
   c. how shells have changed over time.
   d. the age of the oldest star
   Answer: D Difficulty: 1 Section: 1 Objective: 3

7. A hypothesis is
   a. a fact.
   b. a type of question.
   c. a possible answer to a question.
   d. an experiment.
   Answer: C Difficulty: 1 Section: 2 Objective: 1

8. A controlled experiment
   a. tests several different factors at one time.
   b. has several control groups.
   c. has more control groups than experimental groups.
   d. has only one variable.
   Answer: D Difficulty: 1 Section: 2 Objective: 2

Copyright © by Holt, Rinehart and Winston. All rights reserved.

TEST ITEM LISTING continued

9. Which of the following is a limitation of models?
   a. They help explain how something works.
   b. They help describe how something is structured.
   c. They are different from the real things they are modeling.
   d. They are used to make predictions.
   Answer: C   Difficulty: 1   Section: 3   Objective: 2

10. What is the advantage of creating a model of a dinosaur?
    a. Dinosaurs are too large to easily study.
    b. Real dinosaurs cannot be studied because they died out long ago.
    c. Dinosaurs are too complicated to study without a model.
    d. The model is safer to work with.
    Answer: B   Difficulty: 1   Section: 3   Objective: 2

11. A unifying explanation for a broad range of observations, facts, and tested hypotheses is called a
    a. theory.   c. hypothesis.
    b. law.   d. conclusion.
    Answer: A   Difficulty: 1   Section: 3   Objective: 3

12. Life science has few scientific laws because
    a. people don't want them.
    b. life scientists have not done enough experiments.
    c. living organisms are so complex.
    d. scientists need to build more models.
    Answer: C   Difficulty: 1   Section: 3   Objective: 3

13. Many scientific laws
    a. have always been known to be true.
    b. are easily contradicted by new experiments.
    c. often get broken.
    d. may have started off as hypotheses or theories.
    Answer: D   Difficulty: 1   Section: 3   Objective: 3

14. Which technology would be used to view the surface of a tiny living organism?
    a. transmission electron microscope   c. compound light microscope
    b. scanning electron microscope   d. computerized tomography scan
    Answer: B   Difficulty: 1   Section: 4   Objective: 2

15. Scientists use computers to
    a. solve complex equations.   c. produce a three-dimensional image.
    b. produce magnified images.   d. provide chemical protection.
    Answer: A   Difficulty: 1   Section: 4   Objective: 1

16. Which technology would be used to view a person's internal organs?
    a. magnetic resonance imagery   c. electronic computer
    b. scanning electron microscope   d. compound light microscope
    Answer: A   Difficulty: 1   Section: 4   Objective: 2

17. In which of the following areas of study might you find a life scientist at work?
    a. discovering ways to improve computer-operated robots
    b. studying how wasps could be used to control fire ant populations
    c. studying the relationship between El Niño and increased flooding
    d. researching the paths of comets and meteors in space
    Answer: B   Difficulty: 1   Section: 1   Objective: 2

TEST ITEM LISTING continued

18. A scientist who wants to study the possible side effects of a new medicine would probably
    a. give each experimental group the same dose of medicine.
    b. simply ask the subjects about the medicine's effects.
    c. include a control group that received no medicine.
    d. use different numbers of subjects in each treatment group.
    Answer: C   Difficulty: 1   Section: 2   Objective: 2

19. Which tool would a life scientist use to obtain a detailed image of the nerves that branch off a person's spinal cord?
    a. a scanning electron microscope   c. a transmission electron microscope
    b. a compound light microscope   d. an MRI
    Answer: D   Difficulty: 1   Section: 4   Objective: 2

20. An advantage of using the International System of Units is that it
    a. is based on measurements of common body parts.
    b. includes easily understood units of measure, such as inches.
    c. gives scientists a common way to share their results.
    d. uses different units to measure mass and weight.
    Answer: C   Difficulty: 1   Section: 4   Objective: 3

21. In a scientific experiment, a hypothesis that cannot be tested is always considered to be
    a. incorrect.   c. not useful.
    b. illogical.   d. a theory.
    Answer: C   Difficulty: 1   Section: 2   Objective: 1

22. Grams are the most useful unit of measurement in determining the mass of a
    a. truck.   c. grain of rice.
    b. medium-sized apple.   d. hippopotamus.
    Answer: B   Difficulty: 1   Section: 4   Objective: 3

23. A scientific model
    a. can be a kind of hypothesis.
    b. is usually used to represent something simple.
    c. is usually concrete.
    d. is used to explain observations.
    Answer: A   Difficulty: 1   Section: 3   Objective: 1

24. In which of the following areas of study might you find a life scientist at work?
    a. discovering ways to improve computer programs
    b. studying the impact of non-native plants on marshes
    c. trying to develop a warning system for tornadoes
    d. researching the composition of asteroids in space
    Answer: B   Difficulty: 1   Section: 1   Objective: 2

25. A scientist who wants to study the effects of a new fertilizer on plants would probably
    a. give each experimental group the same amount of the fertilizer.
    b. not worry about measuring the amount of fertilizer used.
    c. include a control group that received no fertilizer.
    d. use different numbers of plants in each treatment group.
    Answer: C   Difficulty: 1   Section: 2   Objective: 2

26. Which tool would a life scientist use to obtain a detailed image of the blood vessels in a person's leg?
    a. a scanning electron microscope   c. a transmission electron microscope
    b. a compound light microscope   d. an MRI
    Answer: D   Difficulty: 1   Section: 4   Objective: 2

TEST ITEM LISTING continued

27. What is one advantage of using the International System of Units?
    a. It is based on measurements of common body parts.
    b. It includes easily understood units of measure, such as pounds.
    c. Almost all units are based on the number 10.
    d. It was developed in France.
    Answer: C   Difficulty: 1   Section: 4   Objective: 3

28. A 100 kg object contains the same amount of a matter as a
    a. 1000 g object.
    b. 100,000 g object.
    c. 10,000 g object.
    d. 1,000,000 g object.
    Answer: B   Difficulty: 2   Section: 4   Objective: 3

29. Which work might a life scientist do?
    a. Build robots.
    b. Study Siberian tigers.
    c. Study what affects flooding.
    d. Research comets in space.
    Answer: B   Difficulty: 1   Section: 1   Objective: 2

30. Which tool has a tube with lenses, a stage, and a light?
    a. a scanning electron microscope
    b. a compound light microscope
    c. a transmission electron microscope
    d. an MRI
    Answer: B   Difficulty: 1   Section: 4   Objective: 2

31. Which units are part of the International System of Units?
    a. inches
    b. milliliter
    c. pounds
    d. degrees Fahrenheit
    Answer: B   Difficulty: 1   Section: 4   Objective: 3

32. Which of these is NOT a type of scientific model?
    a. mathematical model
    b. physical model
    c. fashion model
    d. conceptual model
    Answer: C   Difficulty: 1   Section: 3   Objective: 1

33. Which is a step of the scientific methods?
    a. asking questions
    b. stating a theory
    c. using technology
    d. building a microscope
    Answer: A   Difficulty: 1   Section: 2   Objective: 1

34. Which of the following differs between groups in a controlled experiment?
    a. a test
    b. a prediction
    c. a variable
    d. a hypothesis
    Answer: C   Difficulty: 1   Section: 2   Objective: 2

35. Which is a physical model?
    a. a comparison
    b. an equation
    c. a toy airplane
    d. a graph
    Answer: C   Difficulty: 1   Section: 3   Objective: 1

36. What term refers to the amount of space an object takes up?
    a. volume
    b. mass
    c. area
    d. length
    Answer: A   Difficulty: 1   Section: 4   Objective: 3

37. Which describes a compound light microscope?
    a. It passes electrons through something to make a 3-D image.
    b. It passes electrons through something to make a flat image.
    c. It sends electromagnetic waves through the body to make images.
    d. It is made up of three main parts: a tube with lenses, a stage, and a light.
    Answer: D   Difficulty: 1   Section: 4   Objective: 2

TEST ITEM LISTING *continued*

38. Which of the following areas of study is a life scientist most likely to engage in?
    a. determining the density of various plant species in a rainforest
    b. comparing ocean currents in various oceans
    c. finding out how to get rid of a computer virus
    d. studying a volcanic eruption
    Answer: A　　　　　　Difficulty: 1　　　　　　Section: 1　　　　　　Objective: 2

39. In which of the following ways is life science beneficial to living things?
    a. studying rocks on Mars
    b. searching for a cure for AIDS
    c. trying to design earthquake-proof buildings
    d. studying air currents
    Answer: B　　　　　　Difficulty: 1　　　　　　Section: 1　　　　　　Objective: 3

40. Scientists had the following model cells available to them:
    1) a computer-generated model that shows the three-dimensional structures of all cell parts, their relative sizes, and how they interact with each other
    2) a three-dimensional plastic model that shows the relative sizes and the structures of all cell parts
    3) a drawing based on an image from an electron microscope
    4) a three-dimensional clay model that is cut away to show the inside of the cell and relative sizes and the structures of all cell parts
    Which of the following ranking rates the models from the one with the fewest limitations to the one with the most limitations?
    a. 2, 3, 1, 4　　　　　　　　　　c. 2, 1, 3, 4
    b. 1, 2, 4, 3　　　　　　　　　　d. 1, 4, 2, 3
    Answer: B　　　　　　Difficulty: 2　　　　　　Section: 3　　　　　　Objective: 2

41. After many experiments, observations, and testing of hypotheses, scientists hope to develop an explanation for all of them. This explanation would be called a
    a. law　　　　　　　　　　　　c. hypothesis
    b. theory　　　　　　　　　　　d. prediction
    Answer: B　　　　　　Difficulty: 2　　　　　　Section: 3　　　　　　Objective: 3

COMPLETION

42. If life scientists wanted to see the three-dimensional structure of a white blood cell taken from a patient, the most useful tool to use would be _____.
    Answer: a scanning electron microscope
    　　　　　　　　　　　Difficulty: 3　Section: 4　　Objective: 2

43. If a life scientist collects 572 measurements of plants from an experiment, the best tool for analyzing this data would be _____.
    Answer: a computer　Difficulty: 3　　　　Section: 4　　　　Objective: 1

Use the terms from the following list to complete the sentences below.

    mass   electron microscope
    scientific methods   theory
    volume   technology
    hypothesis   compound light microscope

44. Scientists often use experiments to test a _____, which is a possible explanation for observations.
    Answer: hypothesis　Difficulty: 1　　　Section: 2　　　　Objective: 1

45. The _____ of something is defined as the amount of space it occupies.
    Answer: volume　　Difficulty: 1　　　Section: 4　　　　Objective: 3

TEST ITEM LISTING continued

46. A life scientist would use a(n) _____ to magnify a living specimen.
    Answer: compound light microscope
    Difficulty: 1   Section: 4   Objective: 2

47. A set of related hypotheses that are supported by evidence may become accepted as a _____.
    Answer: theory   Difficulty: 1   Section: 3   Objective: 3

48. The use of knowledge, tools, and materials to solve problems and accomplish tasks is known as _____.
    Answer: technology   Difficulty: 1   Section: 4   Objective: 1

49. Using _____, scientists follow steps to answer questions and solve problems.
    Answer: scientific methods
    Difficulty: 1   Section: 2   Objective: X1

Use the terms from the following list to complete the sentences below.

model   scientific methods
life science   controlled experiment

50. A _____ is a representation of an object or a system.
    Answer: model   Difficulty: 1   Section: 3   Objective: 1

51. The steps scientists use to answer questions are called _____.
    Answer: scientific methods
    Difficulty: 1   Section: 2   Objective: 1

52. A _____ tests only one factor at a time.
    Answer: controlled experiment
    Difficulty: 1   Section: 2   Objective: 2

53. The study of living things is _____.
    Answer: life science   Difficulty: 1   Section: 1   Objective: 1

SHORT ANSWER

54. Why is the rejection of a hypothesis helpful to scientists?
    Answer:
    Answers will vary. Sample answer: When a hypothesis is rejected, scientists learn that the hypothesis is not the correct explanation for the problem they studied. They may then seek a different explanation, reexamine data, or ask new questions.
    Difficulty: 2   Section: 2   Objective: 4

55. What is the purpose of the control group in a controlled experiment?
    Answers:
    Answers will vary. Sample answer: The purpose of the control group is to isolate one factor in an experiment so that the factor's effect can be seen. The factor being tested is left unchanged in the control group, but is changed in the variable group. This factor can influence the outcome of the experiment.
    Difficulty: 2   Section: 2   Objective: 2

56. What are the advantages and disadvantages of using a model?
    Answer:
    Answers will vary. Sample answer: The advantages include: they help explain how something works or describe how something is structured; they can represent things that are too small or too large to work with; and they can represent things that no longer exist. The disadvantage is that they are never exactly like the real thing.
    Difficulty: 2   Section: 3   Objective: 2

TEST ITEM LISTING *continued*

57. How could you find the volume of an irregularly shaped pebble?
    Answer:
    Use a graduated cylinder to measure the amount of water that the pebble displaces.
    Difficulty: 2   Section: 4   Objective: 3

58. What is temperature?
    Answer:
    Temperature is a measure of how hot and cold something is.
    Difficulty: 2   Section: 4   Objective: 3

59. If you saw a bird's nest with three eggs nestled in a grassy area, what is one life science question you might ask?
    Answer:
    Sample answer: What kind of bird laid these eggs?
    Difficulty: 2   Section: 1   Objective: 1

60. Who can be a life scientist?
    Answer:
    Anyone from any background can learn to be a life scientist.
    Difficulty: 1   Section: 1   Objective: 2

61. Why is polio no longer a significant health concern?
    Answer:
    Scientists developed a vaccine.
    Difficulty: 2   Section: 1   Objective: 3

62. Why is it important to have a control group when doing an experiment?
    Answer:
    Data from the experimental groups are compared with data from the control group in order to see the effect caused by changes to the variable.
    Difficulty: 2   Section: 2   Objective: 2

63. Why should a hypothesis be testable?
    Answer:
    If a hypothesis is not testable, there is no way to support it or to show it to be wrong.
    Difficulty: 2   Section: 2   Objective: 1

64. How is a scientific theory different from a scientific law?
    Answer:
    A theory is an explanation for a range of data, while a law is simply a summary of a specific set of data.
    Difficulty: 2   Section: 3   Objective: 3

65. What kind of model is a globe, and what are some of its limitations?
    Answer:
    A globe is a physical model of Earth. Its limitations include the fact that it does not support life like Earth and may not have relief features like Earth.
    Difficulty: 3   Section: 3   Objective: 1, 2

66. At one time, life scientists thought that living things could form from nonliving things. Later, experiments proved that living things could only come from other living things of the same kind. What is this evidence of?
    Answer:
    This is evidence that scientific knowledge can change over time as experiments provide new information.
    Difficulty: 3   Section: 2   Objective: 4

TEST ITEM LISTING continued

67. What would be the most useful type of model to show the structure of a protein molecule?
    Answer:
    a physical model
    Difficulty: 2   Section: 3   Objective: 1

68. What would be the most useful type of model to show the idea that the organisms most suited to their environment are the most likely to survive?
    Answer:
    conceptual model
    Difficulty: 2   Section: 3   Objective: 1

69. What cause-and-effect relationship sometimes exists between experimental results and scientific laws?
    Answer:
    If many experimental results all give the same answer, eventually scientists will accept them as evidence of an unchanging scientific law.
    Difficulty: 2   Section: 3   Objective: 3

70. Electrons and electromagnetic waves can be passed through substances to create images. What is the difference in how they are used?
    Answer:
    Electrons are passed through nonliving specimens to create electron microscope images, while electromagnetic waves are passed through living organisms to create either CT scans or MRIs
    Difficulty: 3   Section: 4   Objective: 2

71. A graduated cylinder contains 10 mL of water. A stone is dropped into the cylinder and the water level rises to 17 mL. What is the volume of the stone?
    Answer:
    7 mL or cm^3
    Difficulty: 3   Section: 4   Objective: 3

72. What is the volume of a block of wood that is 3 cm long, 2 cm wide, and 6 cm high?
    Answer:
    $36 cm^3$
    Difficulty: 3   Section: 4   Objective: 3

73. What does temperature measure?
    Answer:
    Temperature is a measure of how hot or cold something is and shows how much energy an object contains.
    Difficulty: 2   Section: 4   Objective: 3

74. What is the SI unit for length? for mass? for temperature?
    Answer:
    meter, kilogram, kelvin or degrees Celsius
    Difficulty: 2   Section: 4   Objective: 3

75. What is the area of a compact disk case with sides measuring 14 cm and 12.5 cm?
    Answer:
    14 cm X 12.5 cm = 175 cm^3
    Difficulty: 3   Section: 4   Objective: 3

76. Give an example of a model that you use in your own life and explain how you use it.
    Answer:
    Sample answer: I use a map to find my way to places I have not been before.
    Difficulty: 2   Section: 3   Objective: 1

TEST ITEM LISTING continued

77. What are two consequences of the work of life scientists?
    Answer:
    Sample answer: Life scientists have discovered ways to prevent diseases such as polio and have discovered how other diseases, such as AIDS, are carried from one person to another. With this information, they hope to find a cure for AIDS.
    Difficulty: 3   Section: 1   Objective: 3

MATCHING

    a. physical model   c. conceptual model
    b. mathematical model

78. ____ a Punnet square
    Answer: B   Difficulty: 1   Section: 3   Objective: 1
79. ____ a toy train
    Answer: A   Difficulty: 1   Section: 3   Objective: 1
80. ____ the idea that life originated from chemicals
    Answer: C   Difficulty: 1   Section: 3   Objective: 1
81. ____ a method of classifying the behavior of animals
    Answer: C   Difficulty: 1   Section: 3   Objective: 1
82. ____ a dinosaur sculpture in a museum
    Answer: A   Difficulty: 1   Section: 3   Objective: 1

    a. area   d. temperature
    b. volume   e. length
    c. mass

83. ____ may be measured in nanometers
    Answer: E   Difficulty: 1   Section: 4   Objective: 3
84. ____ the amount of space something takes up
    Answer: B   Difficulty: 1   Section: 4   Objective: 3
85. ____ a measure of the size of a surface or region
    Answer: A   Difficulty: 1   Section: 4   Objective: 3
86. ____ tells how much energy is in matter
    Answer: D   Difficulty: 1   Section: 4   Objective: 3
87. ____ may be described in grams or kilograms
    Answer: C   Difficulty: 1   Section: 4   Objective: 3
88. ____ SI units are kelvins
    Answer: D   Difficulty: 1   Section: Q4   Objective: 3
89. ____ units for solids are cubic
    Answer: B   Difficulty: 1   Section: 4   Objective: 3

    a. technology   f. mass
    b. volume   g. theory
    c. electron microscope   h. law
    d. scientific methods   i. hypothesis
    e. life science   j. compound light microscope

90. ____ a possible explanation for observations
    Answer: 1   Difficulty: 1   Section: 2   Objective: 1
91. ____ the amount of space something occupies
    Answer: B   Difficulty: 1   Section: 4   Objective: 3
92. ____ used to magnify a living specimen
    Answer: J   Difficulty: 1   Section: 4   Objective: 2

TEST ITEM LISTING continued

93. ____ set of related hypotheses supported by evidence
    Answer: G   Difficulty: 1   Section: 3   Objective: 3
94. ____ the use of knowledge, tools, and materials to solve problems and accomplish tasks
    Answer: A   Difficulty: 1   Section: 4   Objective: 1
95. ____ the study of living things
    Answer: E   Difficulty: 1   Section: 1   Objective: 1
96. ____ a series of steps followed by scientists to solve problems
    Answer: D   Difficulty: 1   Section: 2   Objective: 1
97. ____ used to produce clear and detailed images of nonliving specimens
    Answer: C   Difficulty: 1   Section: 4   Objective: 2
98. ____ the amount of matter in an object
    Answer: F   Difficulty: 1   Section: 4   Objective: 3
99. ____ a summary of many experimental results and observations
    Answer: H   Difficulty: 1   Section: 3   Objective: 3

   a. Ask a question.   d. Analyze the results.
   b. Form a hypothesis.   e. Draw conclusions.
   c. Test the hypothesis.   f. Communicate results

100. ____ Write an article describing what you learned about the ant population on school grounds.
     Answer: F   Difficulty: 1   Section: 2   Objective: 1
101. ____ Last week there were no ants near the front door of our school. Now there is a large colony. Where did the colony come from?
     Answer: A   Difficulty: 1   Section: 2   Objective: 1
102. I think someone released ants from their ant farm near the front door of our school.
     Answer: B   Difficulty: 1   Section: 2   Objective: 1
103. ____ There are three ant colonies on the school grounds. Four of the 10 residents who live near the school also have ant colonies in their yards. None of the residents has ever owned an ant farm. None of the students surveyed had information about where the ants came from.
     Answer: D   Difficulty: 1   Section: 2   Objective: 1
104. ____ Evidence seems to indicate that our rivals, the Hornets, placed the ant colony on our school grounds.
     Answer: E   Difficulty: 1   Section: 2   Objective: 1
105. ____ I am examining the school grounds and surveying students and nearby residents for information about where the ants came from.
     Answer: C   Difficulty: 1   Section: 2   Objective: 1

   a. volume   c. mass
   b. hypothesis   d. theory

106. ____ a possible explanation for observations
     Answer: B   Difficulty: 1   Section: 2   Objective: 3
107. ____ the amount of space something occupies
     Answer: A   Difficulty: 1   Section: 4   Objective: 3
108. ____ the amount of matter in an object
     Answer: C   Difficulty: 1   Section: 4   Objective: 3
109. ____ an explanation that unites a broad range of facts
     Answer: D   Difficulty: 1   Section: 3   Objective: 3

TEST ITEM LISTING continued

    a. life science
    b. technology
    c. scientific methods
    d. law

10. ____ the use of machines to meet human needs
    Answer: B   Difficulty: 1   Section: 4   Objective: 1

11. ____ the study of living things
    Answer: A   Difficulty: 1   Section: 1   Objective: 1

12. ____ a series of steps followed by scientists to solve problems
    Answer: C   Difficulty: 1   Section: 2   Objective: 3

13. ____ a kind of scientific idea that rarely changes
    Answer: D   Difficulty: 1   Section: 3   Objective: 3

ESSAY QUESTIONS

14. A scientist forms the following hypothesis: Classical music improves a person's ability to study. What would you expect to find if the scientist's hypothesis were true?
Answer:
    Answers will vary. Sample answer: If classical music improves a person's ability to study, then on average, people who listen to such music will score higher on a given test than people who study in silence or study while listening to another type of music.
Difficulty: 3   Section: 2   Objective: 4

15. How could you demonstrate that 1 mL and 1 cm^3 represent equal quantities?
Answer:
    Answers will vary. Sample answer: Pour exactly 1 mL of a liquid into a container that has a length of 1 cm, a width of 1 cm, and a height of 1 cm. The liquid should exactly fill the container.
Difficulty: 3   Section: 4   Objective: 3

16. If you want to compare the growth of bean plants in varying amounts of light, what factor will be the variable in the experiment and what factors will you keep constant?
Answer:
    The amount of light will be the variable. Factors that need to be kept constant include the temperature, the type of soil, the amount of water, the type of bean seeds, the number of seeds per pot, and the size of the pot.
Difficulty: 3   Section: 2   Objective: 2

17. Life science students set up the following experiment. They filled four test tubes with the same amount of water and dry yeast, a microscopic organism. They placed each test tube in a different place. Test tube A was in the dark at 50° F. Test tube B was in the light at 60° F. Test tube C was in the dark at 70° F. Test tube D was in the light at 80° F. Evaluate the students' experimental setup.
Answer:
    The experiment has two variables and so the students will not be able to accurately analyze their results. They need to either keep the temperature the same for all the test tubes or the amount of light. The use of the same amount of water and the same amount of yeast meant that these variables were constant.
Difficulty: 3   Section: 2   Objective: 2

TEST ITEM LISTING continued

118. Compare and contrast a transmission electron microscope and a scanning electron microscope.
    Answer:
    Both of these microscopes use tiny particles called electrons to produce clear and detailed magnified images of nonliving materials. The transmission electron microscope produces a flat image and the scanning electron microscope produces a 3-D image.
    Difficulty: 3   Section: 4   Objective: 2

119. Assess the usefulness of the steps of the scientific methods for obtaining information about living things.
    Answer:
    Sample answer: The scientific methods are very useful because they provide a way for life scientists to find out new information and to test ideas. Each step of the scientific methods is important. Without any of them, scientists would not be able to complete their work. A question begins the process and a hypothesis is needed to set up an experiment. Observations are made and recorded, and they are used to form an analysis of results. From these results, scientists draw conclusions and determine if the hypothesis is correct. They share results as a way to get input.
    Difficulty: 3   Section: 2   Objective: 1

120. Explain how scientific knowledge can change over time.
    Answer:
    New evidence resulting from experiments or new technology may contradict an accepted idea, causing scientists to change what they think. Or, the evidence may support an existing hypothesis or theory so that it becomes a law.
    Difficulty: 3   Section: 2   Objective: 4

PROBLEM

Use the table below to answer the following question.

Flowering of Two Plants

	Plant Group	Number of hours of light		
		10	12	16
Average number of flowers per plant	#1 Plant A	0	2	15
	#2 Plant A	0	1	18
	#3 Plant B	10	4	0
	#4 Plant B	14	5	0

121. A scientist collected the following data from an experiment on flowering plants. Analyze the scientist's results and determine what conclusions could be drawn.
    Answer:
    Answers will vary. Sample answer: Plant A produces the most flowers with 16 hours of light and Plant B produces the most flowers with 10 hours of light. Different kinds of plants need different day lengths in order to produce flowers.
    Difficulty: 3   Section: 2   Objective: 3

TEST ITEM LISTING continued

INTERPRETING GRAPHICS

Use the graph below to answer the following question.

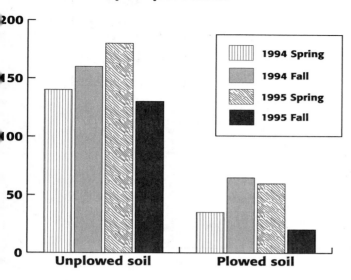

22. What question did the scientists who collected this data want to answer?
    a. Are there more earthworms in the soil in the spring or in the fall?
    b. What is the effect of plowing soil on the number of earthworms?
    c. How is the size of earthworms affected by the seasons?
    d. Does plowing soil affect how fast earthworms grow?
    Answer: B   Difficulty: 3   Section: 2   Objective: 3

23. Where and when were the most earthworms found?
    a. unplowed soil, spring 1995   c. unplowed soil, fall 1995
    b. unplowed soil, fall 1994   d. plowed soil, spring 1994
    Answer: A   Difficulty: 3   Section: 2   Objective: 3

24. What do the data in this graph show?
    a. Unplowed soil has more earthworms than plowed soil.
    b. Plowed soil has more earthworms than unplowed soil.
    c. Plowing of soil has no effect on the number of earthworms.
    d. The number of earthworms cannot be predicted.
    Answer: A   Difficulty: 3   Section: 2   Objective: 3

TEST ITEM LISTING continued

Use the graph below to answer the following question.

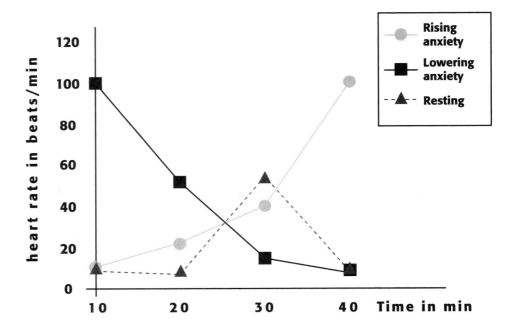

125. The most likely conclusion to be drawn from the experimental results shown in the graph is that
    a. the heart rate is never stable.
    b. only emotions affect the heart rate.
    c. the heart rate changes only when emotions change.
    d. the heart rate increases with rising anxiety.
    Answer: D   Difficulty: 1   Section: 2   Objective: 3

TEST ITEM LISTING continued

CONCEPT MAPPING

126. Use the following terms to complete the concept map below:

controlled experiments
new questions
hypothesis

observations
technology

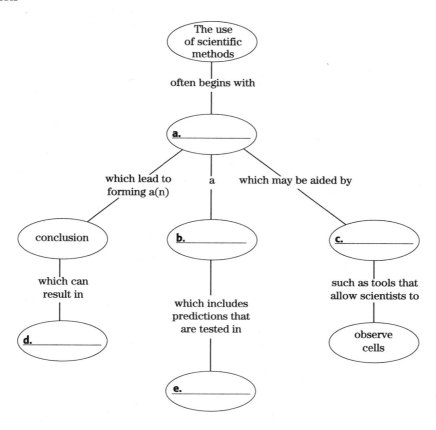

Answer:
a. observations; b. hypothesis; c. technology d. new questions; e. controlled experiments

Difficulty: 3   Section: 2, 4   Objective: 1, 2; 1